BULLETPROOF FEATHERS

EDITED BY
ROBERT ALLEN

BULLETPROOF FEATHERS

How Science Uses Nature's Secrets to Design Cutting-Edge Technology

The University of Chicago Press
Chicago and London

Robert Allen holds a personal chair in biodynamics and control at the Institute of Sound and Vibration Research (ISVR), University of Southampton, UK. He was awarded a PhD at the University of Leeds in the UK for research on modeling the dynamic characteristics of neural receptors. His research interests focus on the development and application of signal processing techniques for biomedical systems analysis and on the bioinspired control of unmanned underwater vehicles.

The University of Chicago Press, Chicago 60637
The University of Chicago Press, Ltd., London

Printed in China
Color origination by Ivy Press Reprographics

19 18 17 16 15 14 13 12 11 10   1 2 3 4 5

ISBN-13: 978-0-226-01470-8 (cloth)
ISBN-10: 0-226-01470-3 (cloth)

Library of Congress Cataloging-in-Publication Data

Bulletproof feathers : how science uses nature's secrets to design cutting-edge technology / edited by Robert Allen.
p. cm.
Includes bibliographical references and index.
ISBN-13: 978-0-226-01470-8 (cloth : alk. paper)
ISBN-10: 0-226-01470-3 (cloth : alk. paper)
1. Biomimetics. 2. Robotics. I. Allen, R. (Robert), 1947–
QP517.B56B86 2010
570.1'5195—dc22

2009037097

(∞) The paper used in this publication meets the minimum requirements of the American National Standard for Information Sciences—Permanence of Paper for Printed Library Materials, ANSI Z39.48-1992.

This book was conceived, designed, and produced by
Ivy Press
210 High Street, Lewes
East Sussex BN7 2NS
United Kingdom
www.ivy-group.co.uk

Creative Director PETER BRIDGEWATER
Publisher JASON HOOK
Editorial Director CAROLINE EARLE
Commissioning Editor VIV CROOT
Senior Editor LORRAINE TURNER
Copy Editors ANDREW KIRK, TOM JACKSON
Art Director MICHAEL WHITEHEAD
Design SIMON GOGGIN, GINNY ZEAL
Illustrator RICHARD PALMER
Publishing Coordinator ANNA STEVENS
Publishing Assistant JAMIE PUMFREY
Glossary ANDREW KIRK
Index CAROLINE ELEY
Picture Manager KATIE GREENWOOD

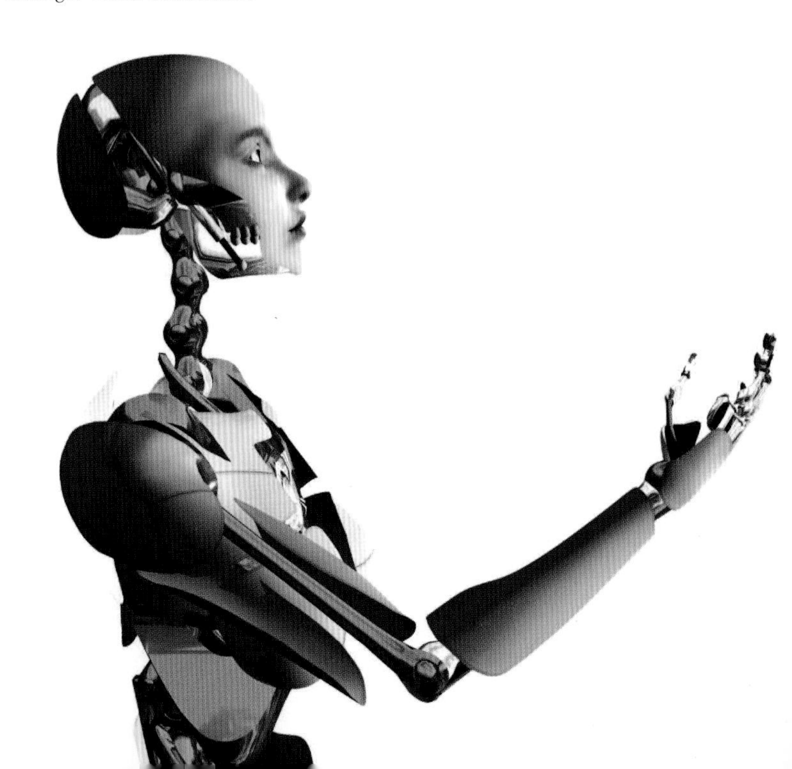

CONTENTS

Introduction

Bulletproof feathers? That's a weird idea, you might think. But imagine if we could develop a bulletproof suit that remained comfortable, light, and flexible until its wearer was threatened, at which point it would harden to offer protection. Not possible? Marine biologists are looking at pointing us in the direction of just this type of adaptable material, developed through natural evolutionary processes over many millennia. From feathers and fins to muscles and mussels, biology can hold important lessons for engineering and technology. How does a barn owl fly so quietly? How can dolphins communicate in the depths of the ocean? Can we apply the principles of human muscle action to create more maneuverable robots? Could a journey into a beehive inspire new methods of cooperative working?

These and a host of other fascinating questions are posed and answered in this book, which covers topics from materials and design through to communication and cooperation. By studying biological principles evolved in nature through successive generations, we can learn how to unlock completely new ways of dealing with engineering problems.

▶ Barn owls are renowned for their ability to fly silently, and an understanding of the characteristics and mechanisms underlying this capability might eventually be utilized in aerodynamics and lead to new innovations in wing design in airplanes.

NATURE VS. TECHNOLOGY

Nature finds the most economical ways to achieve its objectives in terms of energy cost and materials. Biology frequently uses a few simple materials arranged in a particular manner, and makes use of passive methods of sensing or control. Engineering, on the other hand, uses much more energy and often achieves less impressive results. This is not surprising, because evolutionary forces—the survival of the fittest—have operated over millennia, whereas engineering, from the formal perspective of using physical principles and mathematical analysis allied to artistic and functional design, is a relatively modern practice. Of course, humankind has always engineered solutions to problems, but so has biology. We now have the investigative tools of measurement and imaging (down to nano level and below) to be able to study how nature has evolved solutions to the problems it faces. Insights into these mechanisms can inspire and inform human engineering. Research across traditional academic subject boundaries is rapidly increasing and is leading to innovative approaches to a whole host of design problems in areas as diverse as architecture and signal processing.

▼ The fanlike antennae of this male cockchafer beetle (*Melolontha melolontha*) are extremely sensitive and could help inspire new robotic guidance systems.

DEFINING TERMS

The operating envelope of an engineering system is frequently much wider than that of a biological system, which has often evolved to deal with one very specific task. For example, an insect may have antennae with sensors tuned to a male or female call frequency, or to alert it to the movements of a predator, whereas a robotic guidance system might have to operate over a much wider band of frequencies. There are two terms that are used to describe the ways in which biological and engineering systems can be related. One is "bioinspiration," which describes how we can gain inspiration from nature as a basis for developing engineering solutions to problems. This requires an understanding of the natural system and the operational envelope within which it operates. The other is "biomimetics," which is generally taken to mean mimicking nature in some way. The two terms are, however, frequently interchanged, and the newcomer to the subject may become confused. The definitions given here are the most common, and this is the way that these terms are used in this book. It will be seen that inspiration is dominant.

◀ Natterer's bat (*Myotis nattereri*) drinking from a woodland pool. Scientists are studying the amazing ability of bats to use echolocation (sound and echoes) in order to "see" in the dark.

SIGHT AND SOUND

An excellent example of how we might gain inspiration from the biological world in order to develop novel methods of approaching traditional engineering problems is in the area of echolocation. Some animals use acoustics to a very highly developed level. Bats and dolphins, for example, use echolocation to navigate and to locate prey in the darkness of night or the murky depths of the ocean. The performance of humankind's engineering systems comes nowhere near to the impressive feats of these animals. A bat can locate targets in the dark extremely accurately, and dolphins have been shown to use sound to discriminate between different materials. If we could understand the principles of echolocation, then perhaps we could develop improved sonar ranging and imaging systems, geological mineral detection tools, or much-enhanced medical ultrasound imaging systems. Robots deployed to look for surface defects within, say, a nuclear reactor where vision might be impossible or very limited could benefit from acoustic guidance systems. Current technology, however, is very limited and much can be learned from the localization abilities of echolocating animals, which may lead to significant improvements in capability.

▲ Learning how bats navigate and locate targets through acoustics has helped scientists to develop Robocane, which could help vision-impaired people to navigate and pinpoint locations of objects.

FINDING YOUR WAY

Using acoustics to navigate may be of considerable help to someone who has lost their eyesight or has very limited vision. An example of how this can be achieved is Robocane, which is an alternative to the traditional "white stick" that a vision-impaired person might use. The end of the cane holds an ultrasonic transducer that emits sound at frequencies above the human auditory range, which therefore cannot be heard. It also has a receiver that detects the echo signals and transmits the information to the user's hand. Using a tactile feedback system overcomes the problem of giving an auditory signal that might interrupt other important auditory cues such as traffic noise or someone offering spoken guidance instructions. Clearly developments such as this are not directly imitating the abilities of a bat, but they have distilled the principles of echolocation and applied them over a wider operational envelope. Understanding the echolocation process requires considerable input from engineers and physicists with expertise in fundamental acoustics, signal processing, and transducer design, coupled with the specialist expertise of a bat biologist.

▲ Wings of many modern airplanes incorporate fins that were developed as a result of studying the "fingers" on the wingtips of birds of prey. The fins improve airflow over the wings and reduce drag.

GETTING OFF THE GROUND

Flight is another area of human endeavor that shows the influence of ideas derived from nature. Early attempts at flight tried to copy the flapping action of birds and bats; however, despite many elegant wing designs, these efforts were unsuccessful. Now that drag and lift forces are better understood and aerofoil wing sections have been developed to utilize lift forces generated by air velocity differences between the top and bottom surfaces of the wing, flapping flight has largely been discounted by aircraft designers. Birds are experts in controlling lift and drag forces, and we continue to learn from them for fine-tuning modern aircraft wing designs. A barn owl is a master of silent flight and, in common with other birds, has "thumbs" that can be extended into the boundary layer on the upper wing surface to change the airflow characteristics. The small, vertically mounted fins on many modern aircraft wings also developed from observing the effects of the "fingers" on the wingtips of birds of prey, which flex and extend as the bird hovers and glides.

Despite the rapid move away from flapping propulsion in the early days of manned flight, a modern trend of increasing interest is to develop small,

◀ Scientists are studying flapping flight in slender-winged insects such as this southern hawker dragonfly in the hope of incorporating it into more controllable microbots, but flapping flight has largely been discounted for airplane use.

autonomous flapping flight vehicles, particularly for covert surveillance and defense applications. The impressive aerobatics of the dragonfly, for example, have stimulated unmanned air vehicle designers to investigate the possibilities of flapping as a means of hovering, flying backward, and generally maneuvering in tight places. The drive systems to achieve this are, of course, complex and the power requirements are crucial. Nevertheless, the field is advancing rapidly.
As with previous bioinspired developments, the drive to succeed with flapping flight has led to advances in our understanding of insect wings and their effects in flight. The elegant structure of bird feathers is also stimulating aircraft wing designers to look at possibilities for wing surface properties that may lead to improved fuel efficiency and a reduction in noise levels. These investigations deploy advanced fluid dynamic imaging, materials technology, and signal processing, among other engineering techniques, but they also require fundamental biological expertise, and so the two-pronged understanding and development progresses.

GETTING AROUND IN WATER

Like birds, fish and marine mammals make excellent use of lift and drag forces, and offer another potential source of inspiration for the design of underwater vehicle bodies and for novel propulsion systems. A dolphin's skin might inspire the development of a laminar-flow, low-drag hull form, which would be essential for a long-range autonomous underwater vehicle (AUV) if onboard energy sources are to be used most efficiently. A propeller is the usual method of propulsion for a ship or a submarine, but there are numerous alternative propulsion methods employed by turtles, fish, and other animals. Inspired by the flipper actions of the turtle, underwater robotic vehicle designers have developed a flipper-type propulsion unit that allows vehicles to maneuver in small places and even to somersault. The long dorsal fin of some types of pipefish is inspiring designers of underwater propulsion systems to look at oscillating finlike units that would allow an AUV to move forward and backward, and to hover, something that cannot be achieved easily using a propeller. Non-rotating propulsion systems also have a major advantage where there are obstacles that might foul a conventional propeller; for example, when maneuvering through kelp forests or seaweed beds when surveying or tracking marine animals.

Remotely operated underwater vehicles have an array of propeller systems (known as "thrusters") that provide advanced maneuvering capabilities, but they also have the luxury of power supplied down an umbilical line from the surface ship or station. The necessity of carrying an independent power supply really focuses the mind of the designer, and is a major constraint in terms of available propulsion and the duration of mission that can be achieved.

GETTING AROUND ON LAND

Walking robots are on the increase, but control of walking is a complicated process. The animal world frequently makes use of more than two legs, and coordination then becomes extremely complex. Insects, for example, use very elegant strategies for limb control; crabs and other crustaceans have evolved methods of walking in a range of directions that make them extremely maneuverable, even if the main way forward is sideways! Perhaps we can learn from these animals in order to evolve improved strategies for walking machines. If we were designing a walking aid for

▲ The way in which penguins "fly" underwater has inspired the designers of a biomimetic (or "penguin-mimetic") propulsion unit for a kayak. The unit is foot-powered and leaves the arms free for steering or fishing.

someone who had suffered a spinal cord injury, for example, the last thing that person would want is a large backpack housing drive systems, power supplies, and computing equipment. How does an insect control its individual limbs in such a precise and coordinated manner with so few neurons? Could we learn from this to develop control systems using a combination of analog and digital signals? The nervous system uses analog signals extensively in order to integrate inputs from many sense organs and feedback loops, but it uses digital signals (action potentials) for signal transmission over a distance and often in the presence of considerable background "noise" due to the amount of "data traffic" on surrounding neural highways. In engineering, it is now commonplace to control systems via software. Analog signals are frequently digitized near to their source, and integration and processing are carried out in the digital world. Perhaps we could simplify our control system by observing the walking control loop of an insect. Indeed, in the locust, many of the neural connections are well mapped, and the interneurons that link the sensors to the motor neurons are also known. Hence it would be possible to investigate, say, the positioning system of a limb and to determine how the signals are used to facilitate walking. At the same time, the investigative techniques applied can lead to increased understanding of the biological control system.

HOW DO THEY DO THAT?

There are hundreds of other examples of biological processes and techniques that have excited the interest of engineers eager to apply these principles to human technology. The gecko's foot, for example, has attracted considerable attention in recent years. Geckos are frequently seen walking upside down across a ceiling or

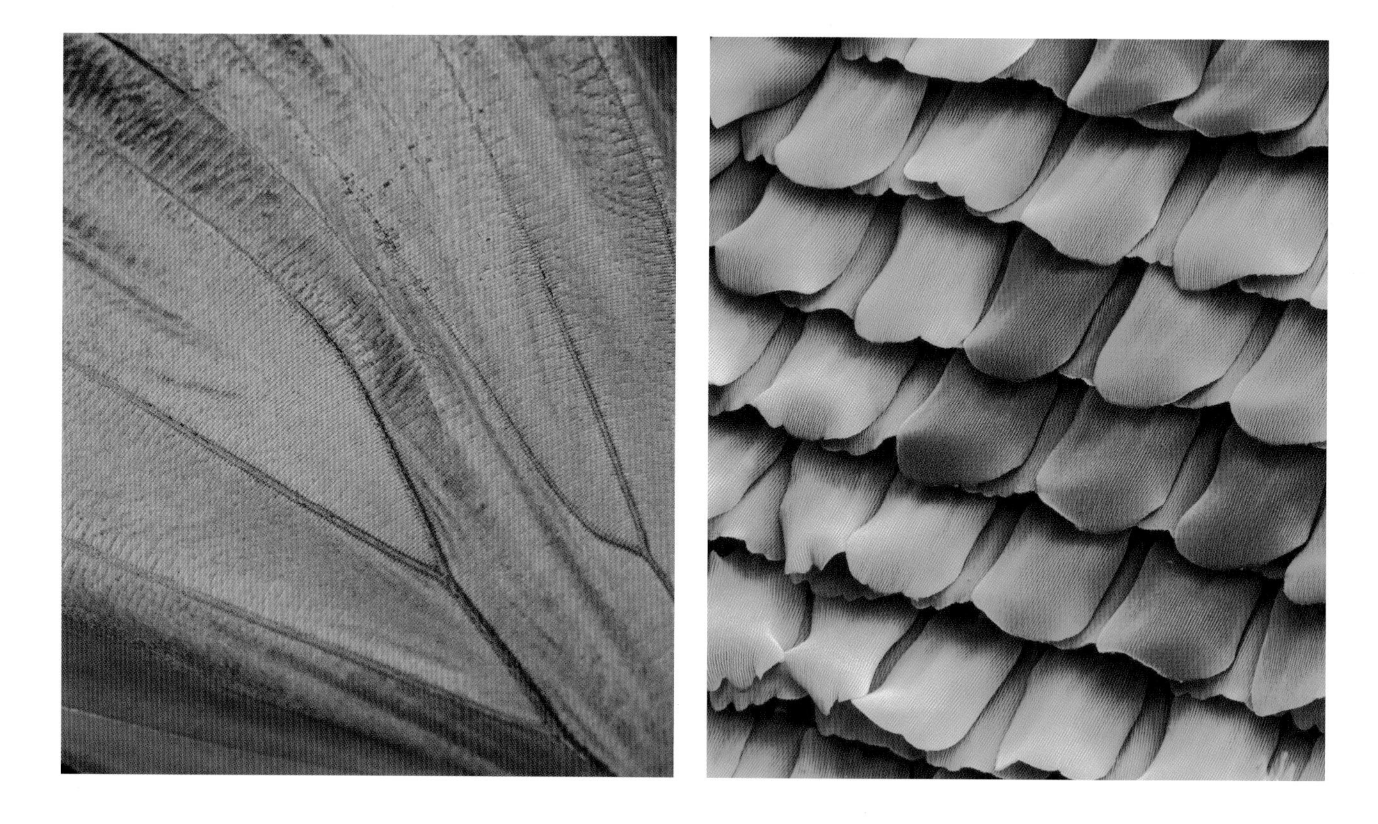

▼ *Below* The way in which butterfly wings glisten with an array of colors has inspired the design of novel display methods in photonics.

▼ *Right* Scales on the wing of a peacock butterfly overlap like roof tiles. They allow heat and light to enter, yet also insulate the insect.

around the globe of a lamp, and the challenge is to understand how these creatures can attach themselves to these smooth vertical surfaces. The understanding gained from these animals is leading to improvements in the adhesive properties of the feet of climbing robots, for example. Thus walking inspection robots could be developed to climb up inside nuclear reactors to carry out material inspection of the internal surface, or to climb the walls and windows of buildings for inspection or surveillance. Antifouling coatings for marine structures are developing from studies of marine plants that are able to keep the surfaces of their leaves or stems clean, where under the same circumstances algae and other forms of marine life frequently foul the hulls of boats and ships. The much-discussed lotus leaf and its ability to repel surface contaminants through the formation of water droplets has also led to developments in self-cleaning coatings for engineering structures. The apparent ease with which a plant's roots force their way through soil is very impressive, and robot designers have been quick to study the mechanisms involved to inform the development of end-effectors for burrowing robots. The list goes on.

A BIOMIMETIC INTRODUCTION

This book is a "primer" on the subject of biomimetics and bioinspiration, based upon the considerable knowledge and experience of some of the world's leading experts. Indeed, since biomimetics is such a young subject, these authors are also shaping the field and pointing toward the future directions of development.

Let us take a brief look inside: marine biology, humanlike robots, underwater bioacoustics, cooperative behavior, moving heat and fluids, and materials and design. This is a diverse range of application areas, but they all share a thread in the economy of materials and processes. They provide excellent examples of how biology and engineering can learn from one another to great advantage.

MYSTERIES OF THE DEEP

Since battery life is limited, underwater propulsion systems are under constant development in order to minimize the energy used, and biology may again hold clues as to how to make improvements. Jeanette Yen describes how some marine animals and organisms have evolved novel ways of recovering energy from vortices generated during swimming. Fish do this by timing their tail strokes at a frequency to maximize this energy recovery. Jellyfish, too, pulse at a frequency that ensures the best energy transfer for propulsion. Engineering systems tend not to capture energy from vortices, but this principle is now being developed for a range of underwater robotic vehicles that aim to cancel out departing vortices so as to generate thrust.

Having a conformable body allows an animal to crawl through narrow gaps, but is little defense against predators. The sea cucumber has developed a way of overcoming this limitation, moving between more and less stiff body states, and we can learn from this to develop mechanically adaptive systems. The field of medical engineering is already developing solutions to the adaptability problem, such as microelectrodes for brain implants that are mechanically "tuned" by their environment. The electrode is initially stiff to enable it to be inserted, but relaxes when it interacts with body fluids in order to comply with movements of the surrounding tissues. Rehabilitation therapy following a stroke or spinal cord injury increasingly uses functional electrical stimulation in which muscles are stimulated by electrical impulses to enable the patient to recover mobility and independence. Surface electrodes are often used, but these require regular removal and replacement. Implanted electrodes overcome this problem, and would no doubt benefit significantly from adaptable mechanical properties, so that in their deformable state they could move with the muscle that is contracting or relaxing.

CATCHING ECHOES

Most ships today have an echo sounder on board to track ocean depth throughout a voyage. The skipper of a fishing vessel uses an echo sounder to look for shoals of fish, but conventional systems use pure tones that are not very useful for classifying targets because the echoes contain limited information. High echo intensity may indicate a large shoal of fish and a much weaker echo a smaller group, but the type and size of individual fish are not discernible. This is also true of seabed mapping, where feature information may be limited. Dolphins, like bats, use a signal with high intensity and short duration. These "clicks" elicit echoes with many frequency components that provide information on the characteristics of the target object such as material composition, thickness, and possible shape. If we could develop engineering systems based on similar principles, this could

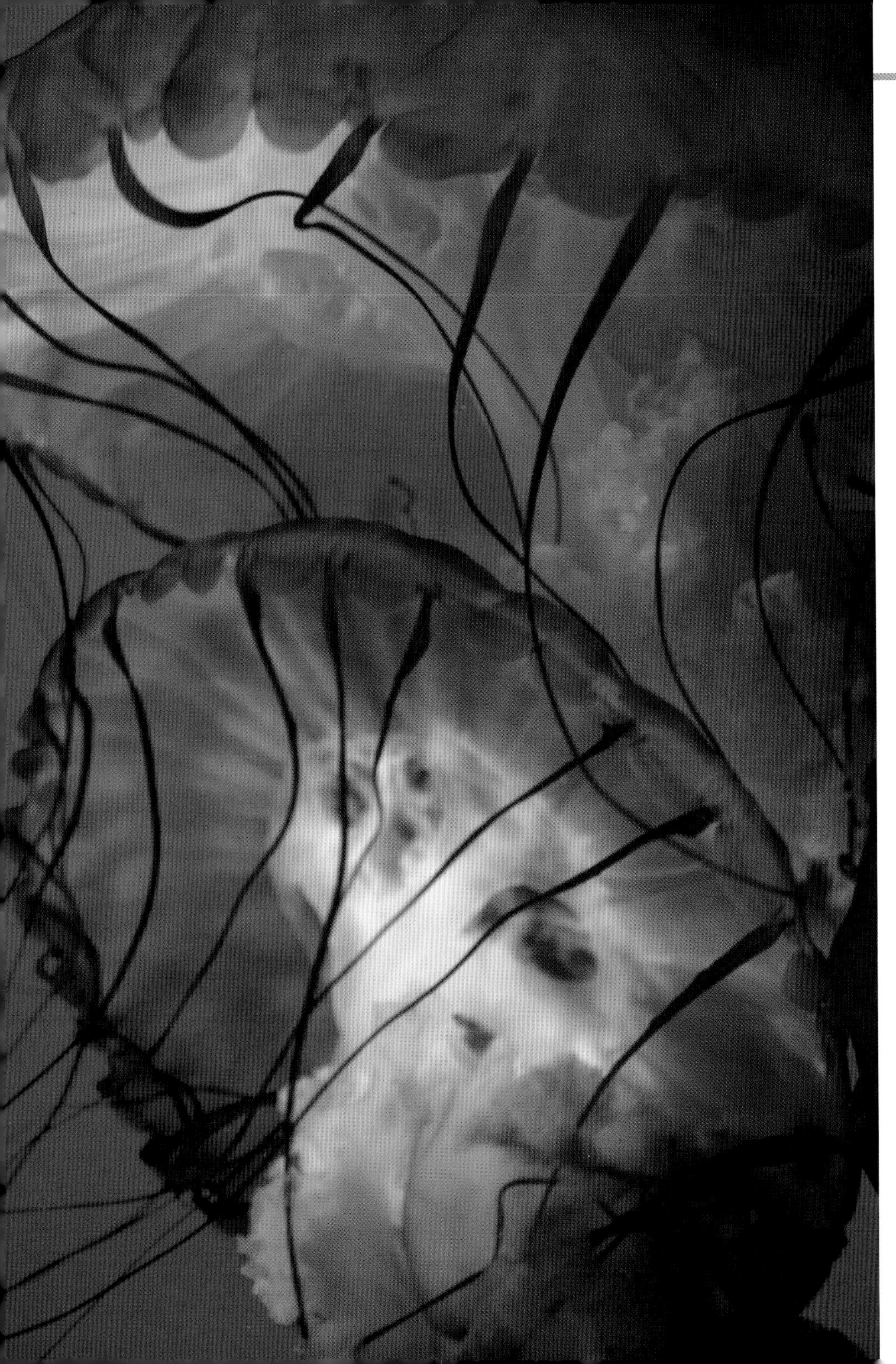

◀ The transfer of energy from the vortices generated by the pulsing action of jellyfish that are used for propulsion may help engineers develop systems to drive underwater vehicles with great efficiency.

mark a significant step forward in underwater exploration, geological surveying, and even medical ultrasound. Tom Akamatsu describes the biosonar of dolphins, which is providing the inspiration for the next generation of echo sounders used to detect shoals of fish. This would offer a means of distinguishing fish destined for the table from those that the fisherman wishes to avoid catching for, say, conservation reasons. If echo sounders were able to classify fish and provide information on the size of a shoal and the size of the fish within the shoal, harvesting could be carried out much more effectively and sustainably.

ROBOTICS

Perhaps the ultimate "biomimetic machine" is the humanoid robot. This may sound like science fiction, but even a cursory glance at Yoseph Bar-Cohen's chapter will quickly dispel such thoughts. Humanoid robots are already here, and Yoseph gives examples in which it can even be difficult to distinguish a clone robot from its inventor. While humanoid robots may have a humanlike appearance, however, it is perception, reasoning, and learning—artificial intelligence (AI)—that will make them really humanlike in terms of behavior. Humanoid robots are much more than gimmicks or technological curios. In space, the deep ocean, or areas of hazard (toxic gases or radioactivity), these robots can perform complex tasks such as repair operations without human agency. The development of anthropomorphic limbs for such robots is necessary to provide dexterity, and these developments can benefit from the work of intelligent prosthetics, and vice versa. In the rehabilitation of patients who have suffered limb amputation, robotic hands and arms are already in use, controlled by electrical signals from muscles, and this exciting area of biomimetic innovation is likely to expand in the future.

WORKING COOPERATIVELY

There is an increasing trend toward small, low-cost robotic vehicles that can be organized into teams, lessening the cost implications of the failure of a single expensive machine. The engineering challenge with these vehicles is to get them to collaborate and work together to achieve their mission. Robert Allen describes how fish schooling and birds flocking can inspire the development of rule-based cooperative control schemes that enable a team to navigate without collision.

◀ The coordinated motion of flocking birds or schooling fish can help establish rules for pilot navigation without collision and for computer animation and games.

Finding the mission target—for example, locating the source of a pollutant in a river or estuary—is quite a different problem. For this we require a different type of inspiration, and the nests of social insects are the source. The distributed sensing and consensus decision-making that take place within a hive of honeybees suggest novel ways to look at the cooperative control of robotic vehicles. These insect strategies can also lead to new ways of dealing with problems such as load balancing on Internet servers or machine tools in manufacturing production. The potential applications of these studies are only just beginning to be understood.

BECOMING AWARE OF THE SURROUNDINGS

In order to take advantage of the lessons learned from social insects, we need sensors for our robotic vehicles. The lateral line of a fish is an elegant mechanoreceptor system, providing it with information about movement and vibration in the surrounding water and enabling it to detect predators or prey. The sensor array detects flow and stationary nearby objects, and forms the basis of shoaling by providing information on movements of local teammates. Similar mechanoreceptors are found on the heads of some fish. Blind cavefish, for example, use these sensors to find their way in the dark and avoid bumping into obstacles. Micromechanical systems are already under development for application on AUVs and other robotic vehicles. This creates the possibility of team operation where such information is key to success.

▼ The study and analysis of the foraging methods of insects such as leaf cutters can lead to new optimization techniques for planning and routing systems.

CONSERVING RESOURCES

Since the increasing costs, the resource limitations, and the environmental impact of our energy consumption have now become urgent issues worldwide, few would deny that saving energy is vital. When we consider that a high proportion of our energy usage is devoted to maintaining comfortable domestic temperatures, a fresh look at heating and ventilation is needed. Nature solves the problems of heating, insulation, and ventilation in many novel and innovative ways, and Steve Vogel gives us an insight into a few of these. For example, connecting cold-blooded legs to a warm-blooded body enables wading birds to feed in very cold water without the loss of heat through the legs. How inspiring is that?

◀ Waders can continue to forage happily for their food in very cold water because the heat of their warm-blooded bodies is not lost through their cold-blooded legs.

CONSERVING MATERIALS

Biology has evolved remarkably effective ways of using materials for structures, and Julian Vincent describes a wide range of stimulating examples and guides us in the way in which we might learn from biological systems for materials technology. For example, engineers expend considerable design thought and energy on avoiding problems created by water, but Julian demonstrates the importance of water for structuring organic materials, and shows the ingenious ways that biology has evolved to create adaptive structures that can change from being compliant to being very stiff. Human skin is among the most remarkable of the structures under discussion. In addition to its impressive mechanical properties, it is self-healing. Not only that, but it also adapts fiber orientation when damaged to protect the wound from expanding—built-in damage limitation! Biology can also teach us about recycling. Biological materials have evolved to be naturally recycled, whereas engineering materials tend to use high-energy bonds and consequently require considerable energy in the recycling process.

Nature clearly offers an abundance of lessons and challenges to engineers. This book demonstrates that the rewards can be significant if biologists and engineers work together, and that the lessons to be learned are illuminating in both directions.

1 | MARINE BIOLOGY

Introduction

How often have we gazed out to sea, entranced by the lure of the unexplored? There is a lot to be learned from the design of the underwater world that could have potentially useful applications for our land-based society. But in order to learn the lessons of the deep, we have first to find a way of exploring this alien region.

Our need to breathe air makes it difficult to explore the oceans in the way that Darwin, for example, made bold forays into the jungles and islands of terra firma. One solution is to send robots to be our eyes and ears, and treasure hunters. Often, the design of these robots and their mechanics of locomotion derive inspiration from aquatic life. The unusual life forms that exist in the sea reflect the effects of reduced gravity, since the force of buoyancy reduces the effective weight that anchors landforms. Two primary principles of nature that have been transferred from studies of aquatic life to enhance the utility of underwater robotics are the capture of vortical energy and the coordination of movement through central pattern generators. Robotuna, robojellyfish, and robostriders rely on momentum from vortices generated in hydrodynamic wakes to be efficient swimming machines. Robosnails and robolampreys, robolobsters and robooctopuses all require integrated control systems to accomplish their tasks.

HOW DO THEY DO THAT?

Three puzzles were solved by testing biological principles using bioinspired robots. Gray's paradox said that bluefin tuna did not have enough muscle mass to generate their observed swimming speeds. Similarly, flimsy jellyfish were not strong enough to achieve the jet-propulsive speeds that they demonstrated. Denny's paradox said that infant water striders could not row their legs fast enough to generate capillary waves capable of propelling them across the air-sea interface. Through separate investigations, scientists came to similar conclusions about how these puzzles are solved. The bluefin tuna flaps its tail at exactly the right frequency to maximize the capture of energy from the previous flap. The jellyfish cinches its circumferential muscles at exactly the right rate to maximize water

◀ The underwater world of the ocean holds many valuable lessons for scientists, which we can apply to our own environment.

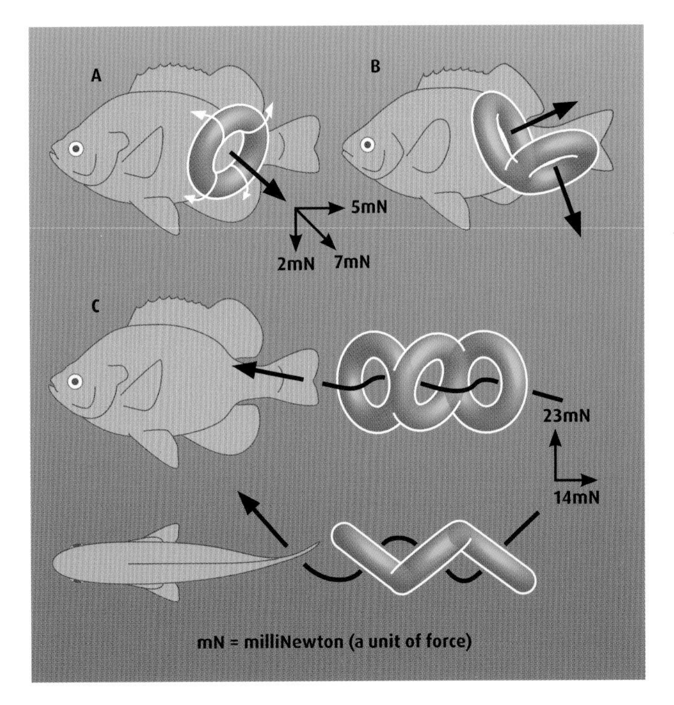

Fish fins shed vortex rings into the wake to produce the force needed for locomotion. A and B: bluegill sunfish and B: black surfperch swimming at 50 percent of their maximal pectoral fin swimming speed; curved arrows represent discrete vortex rings with central high-velocity jet flow (large black arrows). C: lateral and dorsal views of sunfish swimming with the caudal fin, which generates a chain of linked vortex rings in the wake (curved arrows represent central jet flow). Modified from Lauder and Drucker 2002.

transport when a newly created vortex collides with the previous one. For the water strider, thrust from subsurface vortices is generated when rowing, with the water strider's legs serving as oars and its menisci as blades. Momentum from these vortices is translated into fluid motion, to continue the strider's skid across the fluid surface.

All these aquatic organisms capture energy from vortices, an innovation that is extremely rare in human-built devices. Vortices often develop when objects move through a fluid such as air or water. For example, when a wire obstructs an airflow, it leaves a trail of spinning vortices. When a fish flaps its tail, a column of moving fluid that includes thrust-producing vortices is produced in its wake. Following the mechanics of the bluefin tuna, the flap of the tail propelling robotuna was timed to create counterrotating whorls that meet and weaken the ones left by the previous flap. This boost in swimming efficiency solved Gray's paradox. Flapping foils are more efficient than propellers because the force of the foils acts in the direction of motion as a spatial traveling wave along the fish's body. Energy is wasted in propellers, where much of the force is at right angles to the direction of motion. As fish demonstrate, propulsion by flapping foils imparts high maneuverability. The tail moves to generate thrust, leaving the head still, which is useful for sensor equipment.

KEEPING IT SIMPLE

Most interestingly, swimming like a fish does not require complicated control systems or many mechanical elements, just a blade that is moved back and forth. The "trick" that gave rise to lifelike movements in such a simple flapping system is the choice of materials for the tail. The selected materials were able to distribute the hydrodynamic force optimally over the body during propulsion and maneuvering. Also essential is the ability to adjust or tune the frequency of oscillation, to optimize the extraction and recapture of kinetic energy from the unsteady vortical flow with a tuned harmonic drive. The flap of the tail generates a spinning vortex. The next flap of the tail is tuned: its timing is set so that the spinning vortex that it releases into the wake meets vortices of opposite spin. The foil captures the energy of the vortices. It was discovered through experimentation that this specific rhythm maximized the efficiency of the flapping foil for forward movement.

In fact, with this novel capability for vorticity control that produces nonlinear unsteady motion, aquatic swimming organisms are able to outmaneuver any ship or submarine by executing a sudden acceleration. The final design of the robotic tuna distilled the fundamental principles of the natural system and implemented them as well as could be done with current technology and materials. The bioinspired underwater robot served as a test bed for new marine sensors, actuators, and controller technologies.

Harnessing Vortices

When it comes to propulsion, energy, and physical stability, once again the ocean offers us an incomparable treasure trove of ideas and inspiration. Aquatic organisms from jellyfish to water striders show us natural, cheap ways of harnessing vortices, without using vast reserves of energy and materials.

Vortex capture inspired by aquatic organisms lies behind the successful performance of several underwater robots. Propulsion of robojellies by weak contractions of the circumferential polymeric band works only when the pulses are timed such that successive vortices interact with past ones, so that departing vortices generate thrust. Robostriders can walk on water by balancing the surface tension with the momentum gained by vortices created by the rowing of extended limbs.

Passive self-correction of body position in the boxfish also relies on the interacting spiral wake vortices of the body-induced flow. This was the inspiration for the stabilization mechanisms of a bioinspired concept vehicle designed by Mercedes-Benz. The funny little spotted boxy reef fish is an unlikely choice as the model for a car. But speed is not the only function of a car, nor is it that of a fish. A car also needs stability so that it does not flip over. Fluid dynamicists learned from their analysis of the body-induced flow generated by the boxfish that vortex generation was again a key part of the stability solution for self-corrective trimming control. This fish can maintain a smooth swimming trajectory and a stable position, even in a turbulent sea, so that it is able to spin around in a small space such as the crevice of a branching coral.

THE POWER OF THE SEA

Humans have designed devices that harness the energy of motion in oceans and rivers by using the principles of locomotion derived from a careful study of aquatic creatures. A piezoelectric polymeric transducer, inspired by the undulatory movements of the eel, can capture the energy of traveling vortices behind a bluff body when the polymer deforms. This is like a wind turbine converting wind flow into electrical energy. Over time, energy captured from the vortical flow of water by the transducer can be stored in a battery until there is enough power to drive a submerged pump or LED. In contrast to active control, the energy-harvesting "eel" is designed to flap at a frequency best tuned to the local flow dynamics, to maximize strain and the conversion of mechanical flow energy into electrical power. Well-coupled membranes rely on a careful choice of materials that will oscillate easily and lock in, or resonate, at the same frequency as the undisturbed wake. Fish use a similar energy extraction mechanism (getting energy from a vortex) to reduce muscle energy expenditure when swimming behind a river stone or in formation. The key principle learned from nature here is the use of unsteady motion (vortices), rather than steady flow. Utilizing this principle for energy recovery might allow us to exceed the maximum fluid dynamic efficiency of 59.3 percent achievable using a steady flow system, such as that utilized by windmills.

On a larger scale, the Pelamis wave converter extracts energy from waves. *Pelamis* is a Greek word that means "sea serpent." The articulated joints of the Pelamis may have been inspired by the redundant vertebral units of snakes. Unlike swimming snakes, however, which utilize lateral undulatory motion, the movement of the Pelamis harnesses the up-and-down motion of the sea. The freedom we have as human designers is to take inspiration from the biological system and manipulate it for optimal effect. The Pelamis wave converter is made up of cylindrical articulated sections that move up and down as waves pass along it. At each of the hinges between the sections, hydraulic rams use the wave motion to drive generators that produce electricity.

Wave energy is typically more predictable and sustainable than wind energy. The Aguçadoura wave farm off the coast of Portugal, which uses three Pelamis wave converters, is part of an effort to develop renewable energy sources without the carbon dioxide emissions responsible for warming the planet. Aguçadoura, the world's first commercial wave energy project, delivers more than 2 megawatts of electricity, with the intention of meeting the demand of more than 1,500 Portuguese households. Now that's a good way to capture free energy.

◀ **The body-stabilizing ability of the boxfish inspired the design of the boxfish car by Mercedes-Benz** **(see page 163).**

▼ **The Aguçadoura wave farm uses three Pelamis wave converters to help generate renewable energy without warming the planet.**

Control Systems

Locomotion through complex and difficult terrains is a particular challenge under the ocean. Once again, scientists are looking to nature to point the way. This has resulted in a wonderful array of robots, such as robolobster and robosnail, which display some of the amazing capabilities of their natural counterparts.

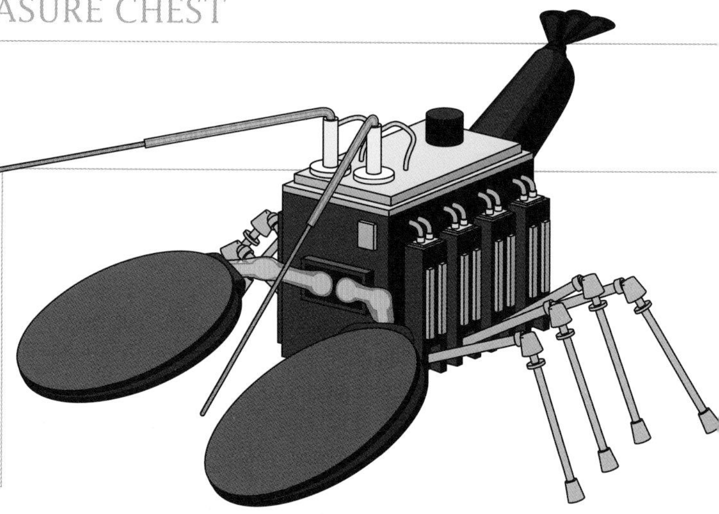

◀ The American lobster (*Homarus americanus*) has helped solve the problem of navigation and adaptation in turbulent fluid environments.

▲ Drawing of an ambulating underwater robot based on the American lobster for use in shallow-water remote-sensing operations.

A key step in assessing the suitability of a natural system for biologically inspired designs is to derive simplifying approaches to use as templates. Robolobster and robosnail illustrate how control of many forms of locomotion can be simplified through clever morphological design and use of functional materials. Crustaceans provide a logical template for underwater robotics. The waterproof exoskeleton, comprised of stiff yet strong chitin plates with an intervening membrane for articulation, protects the interior machinery. The hydrodynamically adaptable shape helps to generate downforce to compensate for the reduced effects of gravity in the aquatic environment. The problem is all those appendages and claws. While they are useful for traction and stabilizing the lobster on the sea bottom, they are difficult to coordinate. One coordination scheme for

▲ This is Robosnail, the robot inspired by the adhesive locomotion of the sea snail, made possible with newly developed soft actuators.

▼ The sea snail has inspired the development of a new kind of robot that can travel in complex environments, even upside down.

autonomous legged underwater vehicles (ALUVs) such as robolobster uses electronically controlled size-ordered recruitment of motor units. By sequentially activating smaller units, then successively larger ones, the resulting movement is smooth rather than jerky. This is useful because sharp movements often catch the wary eyes of predators.

GETTING AROUND

In order to move in the right direction, modular robots can rely on coordination using the principle of a central pattern generator, a simple circuit that can create many different rhythmic patterns by altering the strength and timing of how one element affects the others. Such bioinspired control systems create the backward movement of the robotic marine snail, the traveling wave of the robotic lamprey, and the contractions of the muscular hydrostats that generate the suction of the robotic octopus sucker.

How does the snail provide a good model for robots? Snail movement requires a mucus whose properties change as the stress on the fluid changes. One robotic model of a marine snail, Robosnail I, uses a retrograde wave that propagates in the opposite direction to the snail's motion, a process seen mainly in marine species. Driven by stresses within the fluid layer, the mucus acts like a glue in between waves when the stress upon it is below the critical level, while stresses in the wave region cause the mucus to flow, allowing the advance of the snail. This unusual non-Newtonian fluid enables the snail to anchor the rear end of its body where the mucus is sticky, while pushing the front end of the extensible body forward. Robosnail I has a single foot, making it mechanically simple—like a real snail. Robosnail II has five discrete sliding sections where small motions occur in a coordinated sequence. Like real snails, Robosnail II is stable and is also good at negotiating complex surfaces.

The robotic lamprey also needs coordination to produce the correct wake structure to provide the most thrust, based on the optimal frequency of a two-dimensional pitching airfoil. The many segments of the robotic lamprey must move in a coordinated fashion to create a wave of smooth, seamless motion. Instead of using complex programing to control every movement of the individual segments, a simpler system automatically actuates one section while preventing movement in the other. Modeled on the periodic central pattern generator of swimming slugs, it coordinates the alternate movements of muscles. By adjusting the timing of muscle activity, a traveling wave is generated along the body for eel-like propulsion.

Each robot has a different talent suitable to a specific task. For maneuvering, round robojellies do well. For speed, take the robotuna. To crawl through crevices, try the robosnail or robolamprey outfitted with a flexible robosucker. As implemented in these multitalented bioinspired robots, a guiding principle of the science of biomimetics is to uncover useful natural mechanisms.

Practicing Biologically Inspired Design

How can we anchor a boat in sand? The medium is an unstable mixture of sand and water. The location is distant: at the bottom of the sea, where it is difficult to provide the force needed to dig. Oil rigs have done it, but only with the attendant high cost of extensive manufacturing and materials. What would nature do?

A key question in the practice of biologically inspired design is to reframe the problem by inquiring what processes in nature accomplish similar or inverted functions. We begin by breaking down the problem into functions, and for each function we ask what nature would do. More specifically, how do marine organisms burrow and anchor themselves on the sea floor? This is called "analogical reasoning," a thinking process that can enhance creativity where we take the familiar into the unknown: oil drilling to animal burrowing. The next step is to find champion adaptors—organisms that have faced this challenge over millennia, so that natural selection has eliminated the failures. Will the razor clam and the lowly polychaete worm teach us something we had not thought of before? That is one advantage of biologically inspired design. Instead of using the standard gears and force balances of engineers, we expand our possible design space by understanding how such functional and adaptable organisms use what they have in order to reproduce and transmit that information to their offspring through the blueprint coded in their DNA.

◀ Some polychaete bristle worms have morphologies that are suitable for propagating cracks in mud, thus saving energy when burrowing.

Current technology uses power and the force of pneumatics to drill into difficult materials such as lunar regolith, where some of the soil is like damp beach sand. The drill is a hollow tube of extremely strong material (diamond matrix) that cuts through the regolith, while debris is sucked up the middle of the drill. Isn't that much like excavation by earthworms? In contrast, the polychaete marine worm uses the properties of the environment to save energy when digging in mud. Experimental analysis of the stress patterns created as the worm burrows in seawater gelatin shows that the worm uses an unexpected force when digging: crack propagation. The worm pushes to initiate a crack, which is then propagated by the linear, elastic fracture mechanics of the cohesive mud. The worm drops down into the deeper crack and repeats the process. This is quite different from the force of a drill, and energy savings are gained by passive reliance on such material properties of the environment.

Roboclam, a clam-inspired anchor the size of a jackknife, which replicates clam-digging kinematics with expansion/contraction movements. It has proven energy savings compared to pushing a blunt body through soil.

GETTING A GRIP

What if it is sand that we need to get a grip on underwater? What would nature do here? Check out the razor clam. This clam can dig deeply, down to 27½ inches (70 cm), and exhibits an anchoring force relative to embedding energy that is better than the best anchors. Measurements of the clam's strength show that it is too weak to achieve this, so how does the razor clam perform this function? Analysis of sand movement around a razor clam shows again that the organism has an ingenious interaction with its surrounding environment. The contraction of the flap of the razor clam's shell and vertical body movement fluidize the water–sand mixture. The clam drops its flexible foot or mantle into the quicksand, expands the end of its foot into a bulb, and pulls the shell down behind it. Compared to the energy expended by a blunt body penetrating the soil, energy savings can be 10–100 times. How and what aspects of this biological process can be translated into an engineered design?

Voilà: roboclam! No larger than a Swiss army knife, the roboclam was designed to generate compact, lightweight, low-energy, reversible, and dynamic burrowing and anchoring systems for use in subsea applications. The mechanism utilizes the principles learned from the study of the razor clam—to use the medium to your advantage and reduce the energy required through clever applications of force that result in the same outcome.

Oil rigs use extremely costly digging and anchoring systems that marine creatures such as the polychaete worm and the razor clam employ naturally.

Finding Your Way Around

The intensity of light, as well as color saturation, fades with increased depth in the ocean, as any snorkeler or scuba diver knows. To accommodate these changes in visual information, aquatic organisms have developed unusual perceptive systems—as well as tactics that foil the efforts of one creature to detect another.

The brittle star uses photonic crystals of calcite as light guides, with perfect lenses that focus on sensory cells that inform the star that the shadow of a potential predator is crossing, triggering it to pull and push its flexible arms to move it into a safer hiding place. The mechanism of production of these lenses inspired the nanotemplate for photonic technology. To gather more light in the reduced intensity environment of the deep sea, the stalked eye of the lobster is endowed with multiple lenses focused on one

◀ The lobster's eye has panoramic vision all around itself, and has inspired scientists to create a new breed of image detectors.

▲ The Lobster-ISS, mounted on the International Space Station, has six detector modules, each pointing in a different direction.

◀ The cuttlefish's amazing ability to adopt striking different colors has led to new innovations in brighter display technology.

or more photoreceptors, thus amplifying the signal input. This compound eye structure gives it panoramic vision, inspiring the design of an all-sky telescope to register small movements of celestial bodies in the night sky. For organisms with large lenses, as found in mammalian eyes, the water around them serves to repair aberrations in spherical lenses. To repair the eyesight of humans, we now have goggles with adjustable lens chambers filled with refractive fluids—adaptable spectacles for people in areas lacking in optometrists.

NOW YOU SEE ME

To outwit predators that are able to amplify their visual system, prey in the sea deploy camouflage schemes. The cuttlefish can disappear before our eyes, by changing its shape and color. A series of color layers provides the cuttlefish with a spectacular palette. The surface layer contains chromatophores where pigment colors change with skin depth, starting from yellow at the surface, going to red, then brown when deeper under the skin. Each color is a separate chromatophore. Neurons control the size and shape of these pigment cells. When the cell expands, the color is seen; when it condenses, the color disappears. Adjacent to the chromatophores are flat platelets called "iridophores," with iridosomes oriented on edge that produce iridescent metallic green or blue colors, enhancing the brilliance of the overlying pigments. Below this layer in the dermis are the leukophores. These contain translucent reflective protein granules of high refractive index that reflect all wavelengths and thus act as a broadband reflector. White is reflected white and blue as blue, thereby enabling the cuttlefish to blend into its surroundings.

A key property of cephalopods (such as cuttlefish and octopuses) is the spacing between protein layers. These nanometric variations in spacing are equivalent to the wavelengths of light and thus have the ability to bend light, enabling color change. Phototonic sensor designs are taking inspiration from the nano-spaced structures of these protein layers: human-made optical gas sensors detect vapors that fill the nanometer gaps between layers, resulting in a color change associated with that chemical solvent or gas. Light is both reflected by and transmitted through the layers, making many colors possible. Color shifting by cuttlefish has inspired the mechanism used to generate brighter display technology, where spacing can be adjusted by applying voltage. The ease with which colors can be tuned by varying the voltage opens up a wide array of applications for this bioinspired invention.

Feel the Flow

Since the availability of useful light diminishes with depth, light-based perception may not be the most useful sensory modality for roaming the high seas. As a result, animals that live in the ocean depths have developed other methods for establishing what is happening around them.

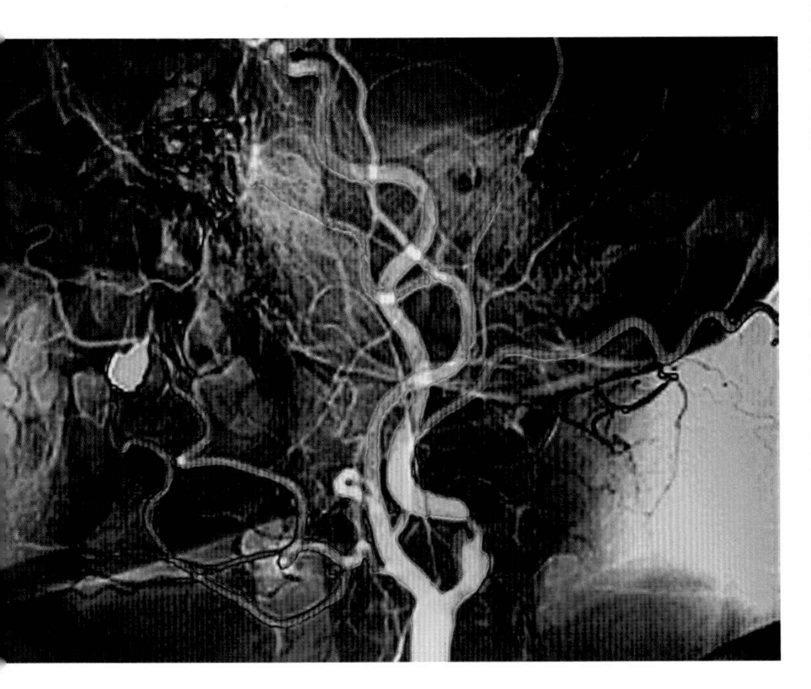

Studies of flow detection show that undersea organisms have mechano-sensory arrays that are ultra-sensitive to fine-scale flow. Delicate but chitinous setal hairs along the long antennules of the tiny microcrustacean copepod are separated by only tens of micrometers. Such closely spaced sensors are capable of comparing extremely small-scale velocity gradients. Tiny perturbations in flow like these occur around plaque deposits in blood vessels, so a potential medical application of this biotechnology would be a sensing microrobot that detected these bumps and leaks, and released plaque-dissolving drugs on the spot. The exquisite lateral line or neuromast of a fish has a sensitivity similar to that of the copepod, and this sensor has been emulated technologically.

The copepod relies on the viscosity of the fluid to smooth out small-scale turbulence, thus filtering out noise and background interference. The fish is larger and moves faster and it finds another solution. The hairlike sensors, similar in size and scale to the setal sensors on the copepod, are embedded in a fluid-filled canal, the fish's lateral line. Furthermore, the hairs of the neuromast—the collection of cilia that form the individual flow sensors of fish—are covered by a gel-like matrix or cupula that filters high-frequency flows that are not representative of prey, predator, or mate. In the construction of an artificial flow

◀ **The plaque deposits in this neck artery could one day be detected by medical equipment based on the copepod's hairlike sensors.**

▲ **Minute water flows are detected by the fine hairs that adorn the first antennae of that much-studied microcrustacean, the copepod.**

▼ Blue crabs have odor sensors on their feet and antennae, which enable them to track odor plumes.

▶ Fast-swimming fish embed their neuromasts (flow sensors) within the canal system of the lateral line.

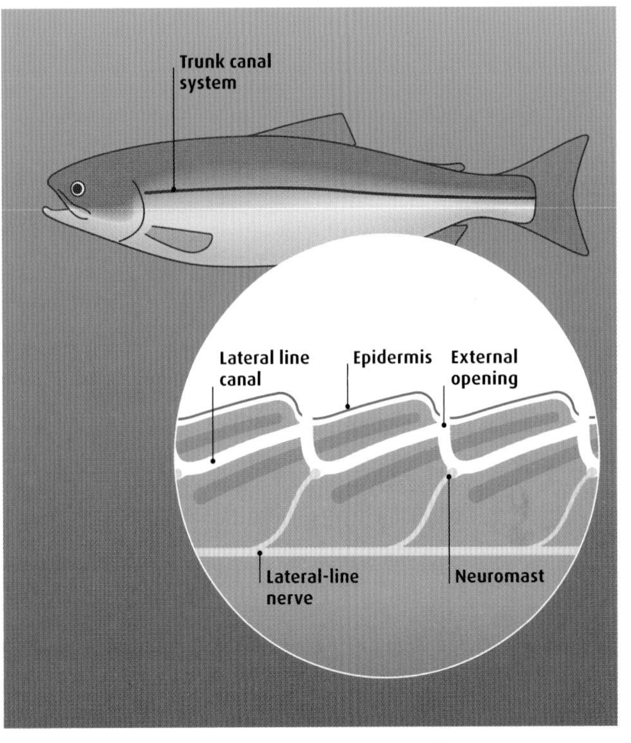

sensor for self-navigating automonous underwater vehicles, a glycoprotein hydrogel with similar material properties to the fish cupula was grown. This increased the drag of the artificial cupula, causing movement at much smaller flow speeds and amplifying the signal, extending the source-to-sensor distance.

FOLLOWING THE SCENT

Another key sensory modality underwater is odor detection, and here we look to the blue crab as our biological inspiration. The crab offers an ingenious model of odor plume tracking. It is equipped with sensors on its antennae and legs that broaden its spatial coverage. The key advantage of this distributed sensory system is that the antennae samples above the boundary layer to detect strong signals, while the feet sense odors on the bottom, where a homogeneous signal allows more integrated assessment of the plume boundary.

Scientists have used the behavioral strategies of animals as a way to organize the program architecture of artificial algorithms used by robots to follow odor plumes. For instance, the computer program used to track aquatic chemical plumes causes the robot to cast back and forth across the flow in search of odor not only at the beginning of the run, but also when the robot has not detected odor for some period of time. The switch between plume following and casting is a common feature of animals that track turbulent chemical plumes, and is a way to deal with the intermittent structure of these signals. Long periods of signal absence may indicate the searcher has lost contact, resulting in a switch from moving upstream in response to odor to moving cross-stream in an effort to find the plume. The strategies employed by animals have been honed by evolution to deal effectively with signal variability. Knowing what the animal does has resulted in algorithms that can work even in these unpredictable circumstances.

Immovable Creatures

Millions of years ago, when terrestrial predators exerted their selective pressures on their land-bound prey, a few organisms got away and opened up a new niche. These were the geckos, flies, spiders, and beetles, which developed extraordinary abilities to adhere to a variety of surfaces.

The gecko sticks nearly effortlessly to smooth rocks and overhanging branches, over and over again, with no loss of sticking power. Its feet remain clean. How?

Analyses of the nanostructures of lizards and insects reveal a simple but unexpected solution. Through convergent evolution, spatula-like nanostructures enabled these light- to heavyweight organisms to adhere to surfaces. Simply by subdividing the tiny hairlike setae to maximize surface density, it was discovered that geometry is the central design principle in the evolution of adhesive setae. Adhesive force is proportional to the linear dimension of the contact. Instead of synthesizing specialized surface chemistry to coat epidermal structures, lizards and insects alike rely on subatomic energies called "van der Waals forces," which are created by the keratinous setae in lizards

◀ The gecko's clean feet are able to stick to any smooth surface through the use of dense hairlike setae.

◀ Bearded mussels utilize adhesive proteins that enable them to adhere to slippery submerged rocks in the ocean.

or the chitinous setae in insects. For example, for a fly puvillus (a hairy adhesive organ at the end of a fly's leg), an order of 103 to 104 setae is needed to provide the force to support the insect. By increasing the number of hairs, larger organisms can utilize the adhesive properties and climb up to escape from land-bound predators. This elegant example of hierarchical engineering repeats one of biology's key lessons: the reliance on the massive summation of minute forces to achieve a macroscopic result.

ADHESION IN WET ENVIRONMENTS

A similar evolutionary challenge has been met in the sea. Bearded mussels cling to the edges of slippery submerged rocks. Few predators could withstand the force of ocean waves to capture the tasty bivalves. What keeps these mollusks from being lost out to sea and how did these invertebrates manage to stick one wet surface to another? There are no suckers like the ones on an octopus, nor are there crablike claws. Instead, these spineless creatures use a novel adhesive characterized as an L-DOPA and hydroxyproline composite of proteins. These adhesive proteins are secreted by the foot organ into the byssal groove, to form the thread that attaches the black mussel *Mytilus edulis* to the rocks on wave-swept coasts.

Using this template for byssal thread and plaque formation, the reactive oxidized form of DOPA, quinone, is thought to provide the moisture-resistant property of mussel underwater adhesion. The attached thread at the distal connection to the rock is as strong as vertebrate tendon but three to five times more extensible. The –800 pN (pressure rating) of force to dissociate DOPA from an oxide surface was the highest ever observed for a reversible interaction between a small molecule and a surface.

GECKEL: COMBINING DRY AND WET ADHESION PROPERTIES

Can humans learn from these ingenious solutions that have worked so well for geckos and mussels over millions of years? Indeed, in 2002, a biomedical engineer called Phillip Messersmith put two and two together. The tetrapod and the bivalve provided the basis for a compound solution, something that nature could not do. Nature cannot easily combine the traits from unrelated animal groups. Humans were able to do this because we understood the principles. By merging the dry adhesive properties of the gecko with the wet adhesive properties of the mussel, we produced "geckel"—a strong yet reversible wet/dry adhesion with a property that existing materials do not exhibit. This required quite a bit of nanopatterning for the polymeric footprint and quite a bit of chemical engineering for the layer of reversible glue on the tips. The result was a bioinspired design that enables surgeons to adhere the two wet sides of a cut tissue.

For flexibility to adapt to rough surfaces, poly (dimethyl-siloxane) (PDMS) elastomer was used in the micro-fabrication of the pillar arrays to mimic the gecko's foot hairs. This nanostructured surface was found to be essential to the geckel adhesive behavior. The pillars are coated with a mussel-adhesive protein-mimetic polymer topped with an organic layer of catechols, a key component of wet adhesive proteins found in mussel holdfasts. This increased the adhesion by nearly fifteenfold and the system proved able to maintain its adhesive performance for more than a thousand contact cycles in both wet and dry environments.

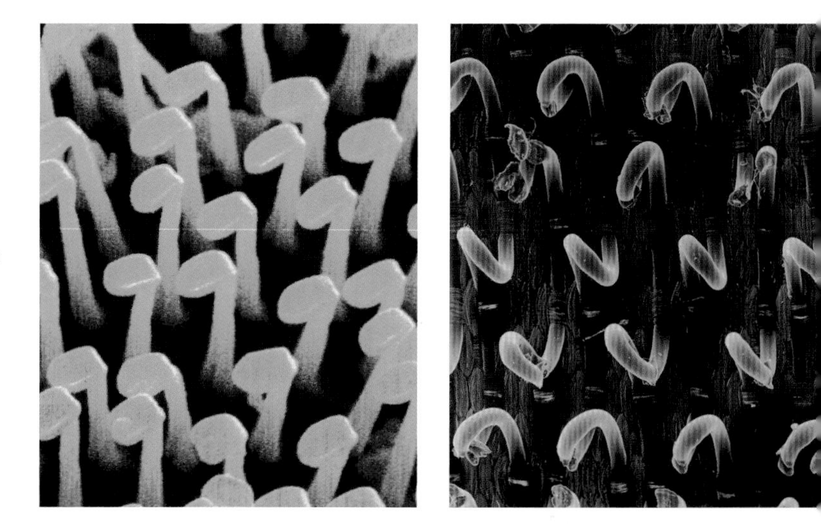

▲ The blue hooklike structures above are the adhesive setae (hairs) on the end of a fly's foot. This phenomenally adhesive structure is similar to the structure of the clingy seedlike burr that inspired scientist George de Mestral to invent the material on the right, which has become known the world over as Velcro®.

Squeaky Clean

Have you ever seen a dirty dolphin or a filthy coral? Perhaps the dolphin can roll around in the sand to rid itself of hitchhikers, but corals cannot move like that. How do these aquatic creatures keep their skin so clear of debris? What can we learn from the underwater world about keeping surfaces clean?

Some crustaceans do indeed shed their encrusted outer layer and replace it with a clean one when they molt in order to grow larger and simultaneously rid themselves of epibionts or surface parasites. Corals can produce copious amounts of energy-costly mucus that deters settling and sloughs off any settlers. Other organisms such as decapods groom their skins to keep them clean. Sharks keep their slick suits clean using microtextures that prevent gametes and colonizers larger than the skin-surface crevices from settling. But it is the clean kelp that has come up with the inspiration for a possible medical breakthrough. Instead of disrupting the settlement of parasites by surface topology or by unveiling a clean, new layer by shedding, the kelp produces a chemical that disrupts the group attractive behavior of biofilm-forming bacteria. The tendency of bacteria to form these films on hospital surfaces, or on the human body, makes them very dangerous, and the kelp supplies an ingenious solution.

STRENGTH IN NUMBERS

Biofilm formation depends on the ability of bacteria to detect each other's presence. These bacteria "talk" to each other using quorum sensing, where chemicals released by neighbors trigger a series of responses in a bacterial cell that are necessary to form biofilms. In this way, a particular bacterial cell undergoes these changes only when the surrounding bacterial population is sufficient and appropriate. Yet even in the richness of the sea (a bacterial soup, if there ever was one), few bacterial films trouble the red alga *Delisea pulchra. Delisea pulchra* produces a secondary metabolite that

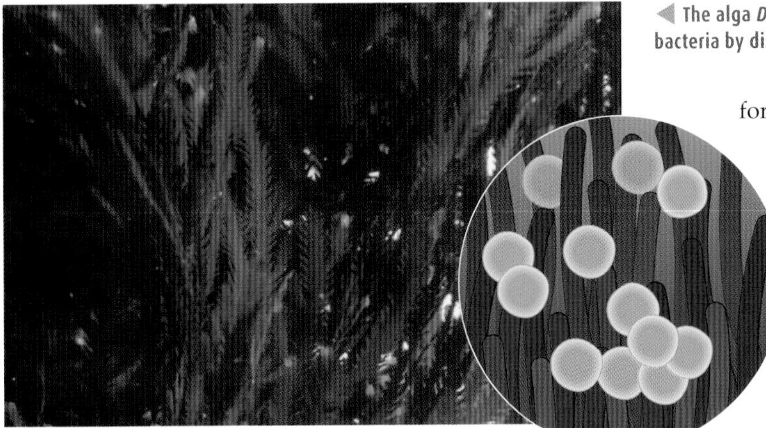

The alga *Delisea pulchra* avoids contamination from bacteria by distorting the messages they send.

for quorum sensing. The bacteria cannot ignore this message because it is the same one they use to form colonies. Hence bacteria use this signal to form colonies on hospital surfaces, but fail to form the film on the kelp. Taking this cue from nature, humankind has a new way to combat bacteria. This allows us to suppress the formation of biofilms without promoting the evolution of resistant bacteria, which has foiled each generation of antibiotic we develop.

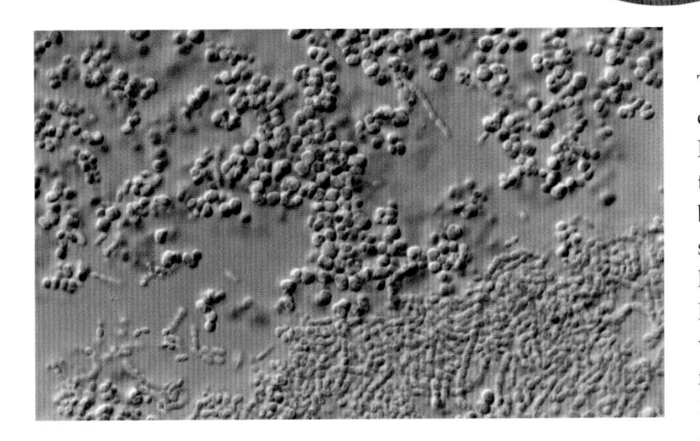

The clean kelp secretes a chemical to discourage parasites and bacteria from forming on it.

Delisea pulchra mimics the biofilm signal shown here, in order to trick and evade bacteria.

inhibits bacterial colonization by interfering with cell signaling. Instead of killing the microbes, these chemicals interfere with the essential message produced by the colonial-living bacteria. The compounds mimic the structure of the biofilm signal, but not the response. Bacteria cannot defend themselves by evolving a resistance to this dishonest message because this would interfere with the bacterium's own survival strategy. Like camouflage, this is another example of the evolution of deceptive signals. The kelp has eavesdropped on the signaling system of the bacterium and mimicked the cue

A STICKY SITUATION?

To continue searching for inspiration from nature, we can consider an opposing function of cleanliness: stickiness. For this, the adhesive system in mussels has been revisited, this time not for its amazing ability to stick to a wet surface, but for the same function of self-cleaning that *Delisea* solved so elegantly. As mentioned earlier, dopamine is the key compound that allows the mussel to stick to rocks. New antifouling coating formulations using dopamine were able to prevent settlement and adhesion of the filmmaking diatom *Navicula perminuta* as well as the alga *Ulva linza*. Borrowing a secret from the mussel, this coating did better than our standard silicone coating in laboratory tests. The challenge is one that is common when attempting to translate biological principles to engineered design that humans can use: can this method be scaled up from the job of keeping those blades of algae clean to the task of keeping the vast surface area of a ship's hull clean? Research into this question is continuing.

Lessons in self-cleaning learned from marine creatures could inspire ways to keep boats and ships clean.

Tailored by the Power of the Sea

Never turn your back on the sea. How often are beachcombers given this warning so that they are not carried off by the power that is packed in a wave? Yet, as beachcombers know very well, many aquatic creatures have found a way to live in the surf and withstand the impact.

Many marine biomaterials are derived from minerals extracted from seawater and arranged in a precise structural alignment. With only biologically compatible ingredients, a number of material properties are made manifest. Similar materials are used for different functions, and heterogeneous materials converge on similar functions. Strong structures are built from the bottom up using locally adaptive building techniques. Is it possible for humankind to match that?

UNDERWATER CHEMISTS

Take $CaCO_3$, the biomineral of abalone and coral. To obtain this from seawater, corals pump protons out of a small calcifying space initially filled with seawater, raising the pH of the seawater in that space. This converts the dissolved inorganic carbon in the calcifying space to carbonate ion, increasing the saturation state ($Ca \times CO_3$ ion product). Once the saturation state is raised enough, aragonite precipitates. The abalone creates an impact-resistant shell of 95 percent calcium carbonate "tiles." The remaining 5 percent is a

▶ The biosilica spines of this glass sponge channel light and have the same design as those used in fiber-optic telecommunication.

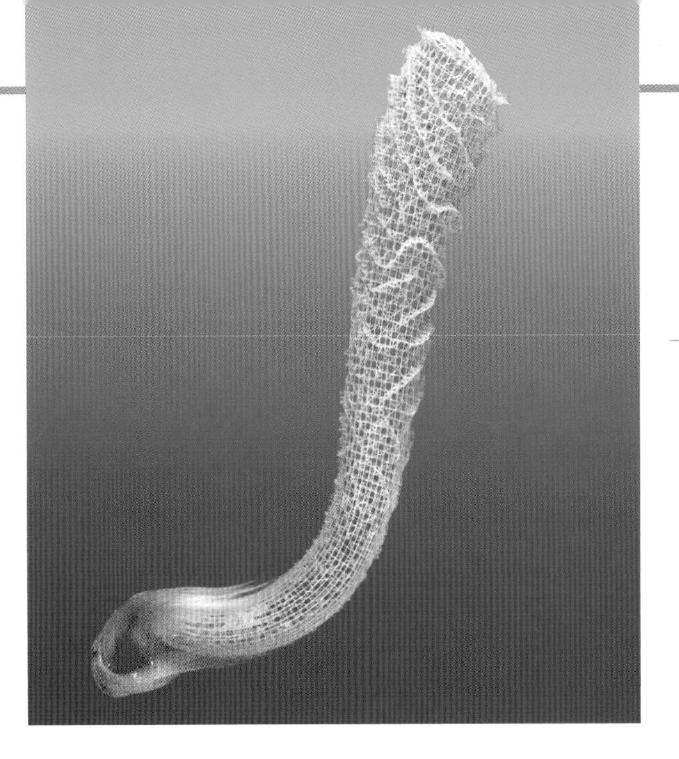

protein adhesive. Impact forces dissipate along the layers, crack propagation is limited by the offset tiles, and cracks are healed by filling in the fissures. (Materials scientists study these principles to develop lightweight body armor.)

SMALL-SCALE MIRACLES

Using the same material, calcite, the brittle star makes an optically perfect biolens. All over its flexible body, the star has double lenses placed in an array designed to prevent spherical aberration and birefringence (the splitting of a ray of light into two rays). Such biocrystallization has been used in micropatterning processes. In glass sponges, light guides are made of biosilica spicules or spines. The spine has the same design that is used in fiber-optic telecommunication: cladding around an inner core. Here the core has a high-refractive-index material comprised of sodium-doped silica and a cladding with a lower refractive index. The silica-based waveguide channels light inside the fiber, and it is confined by the outer surface. Silica spines provide the potential ability to perceive light from different directions, and structural analyses show seven hierarchical levels of sponge glass over seven orders of magnitude (0.000001969–20 inches/ 50 nm–50 cm). The laminated structure and interspaced organic material promote crack deviation. These features impart high mechanical stability so that, when arranged in a specific framework, the glass spines provide increased fracture resistance to the forces that the sponge encounters.

Diatoms mineralize biosilica and are considered one of nature's most gifted nanotechnologists. They are tiny single-celled organisms that produce 20 percent of the world's organic carbon through photosynthesis. The glassy cell wall of diatoms is a strong lightweight mechanical structure that results from the presence and arrangement of amorphous silica. Through genetic engineering, efforts are underway to create diatoms with novel silica structures for the synthesis of nanostructured inorganic materials. Other researchers utilize the siliceous diatoms themselves as lithographic templates for intricate nanopatterns. Shape-preserving reactions, where the silica is displaced by other materials, such as magnesium, transform their nanostructure to enable the mass production of nanodevices for use as sensors, filters, and optical gratings. All these ideas for impact protection, nanopatterning, and light capture are offered by the sea.

◀ The biomineral of coral is aragonite ($CaCO_3$—see inset). This tough but lightweight mineral is being studied to mimic its impact-resistant properties.

▶ Fiber-optic lines are made of optically pure glass and can be as thin as human hair. They can transmit light waves over very long distances.

Hard and Soft

A firm structure is often a useful protection, but it can be inflexible and difficult to maneuver. Sometimes it is better to be soft. But it is better still to have both states at your disposal. Some marine creatures, such as the sea cucumber and hagfish, have found a way of achieving just that.

The sea cucumber can crawl between narrow spaces when soft, but when threatened by predators it undergoes a phase transition with the addition of water. The sea cucumber's skin is composed of an ultrafine network of cellulose fibers, or "whiskers." In defensive mode, surrounding cells release molecules that cause the whiskers to bind together, forming a rigid shield. The tightness of the connections between the collagen fibers determines how stiff the cucumber's skin is, and is

◀ The sea cucumber is soft enough to crawl into narrow spaces, but can transform its skin into a rigid shield when under threat.

▲ A bulletproof vest that is soft to wear but can harden during combat may be made possible through study of the sea cucumber.

The hagfish has slime glands that can release large amounts of stringy slime when it is attacked, provoked, or stressed.

controlled by the animal's nervous system. In a relaxed state, other cells release plasticizing proteins, loosening the fibers and making the skin pliable.

Biologically inspired applications include mechanically adaptive microelectrodes for brain implants for conditions such as Parkinson's disease and stroke, and for spinal cord injuries. The implant needs to be stiff when inserted, then relax once it interacts with brain fluids. Scientists have developed a chemoresponsive mechanically adaptable polymeric substance that works like the sea cucumber's skin. The cellulose nanofibers—with the rubbery polymer ethylene oxide–epichlorohydrin—form a stiff network. Due to the nature of the bonds between the polymer and the fibers, water in the cortical tissue can get between the two substances, weakening the fibers' adhesion. The material then becomes soft so that it can mold itself, thus minimizing damage to the brain tissue. Other biologically inspired applications might include the design of a bulletproof vest that can harden when a soldier goes into combat, but be more comfortable when the wearer is out of harm's way.

A SLIMY CUSTOMER

Going even softer, we can take a look at hagfish slime. This is full of intermediate filaments (IFs), intracellular structural proteins that are important components of the cytoskeleton, as well as of skin, hair, fur, nail, claw, hoof, and horn. Hagfish are the only animals known to secrete IFs externally in their defensive slime. The slime is full of water, which slips out when the gooey mass is picked up. This superabsorbent material can absorb 260,000 times its weight in water. The secret lies in the multiple space-occupying filaments: within the secretory gland, the filaments are wound up into tight spindles of fiber that unravel when pulled. Water is retained because the filamentous tangle produces pores that restrict the flow of water, very much like viscosity. While the hagfish's predator suffocates when embedded in the slime, the hagfish can twist its body into a knot and slide this knot down its body, to shed itself of the secreted slime and escape from this viscous mixture of mesh and predator. If IFs can clog a fish's gills, what else could they clog and seal?

Hagfish slime differs from other slimes in that it contains not only mucinlike molecules, but also a fibrous component built from IFs that has an extremely low modulus of elasticity. The tensile properties of hagfish IFs show that they behave like nanoscale rubber bands that can recover fully from strains of up to 35 percent. After this, the helical coils begin to extend and form stable sheets of crystals with neighboring protein strands, which fail only at strains of more than 200 percent, so that they are far stiffer near their breaking point than they are at low strains. The hardness varies according to the amount of water that saturates the slime. IFs have been coopted for use in a huge diversity of biomaterials, including soft cytoplasmic gels, rubbery tensile elements, and stiff and fibrous hard alpha-keratins.

THE BENEFITS OF COOPERATION

Functional analysis of problems and analogical comparison with natural sources are important steps in the process of biologically inspired design, an interdisciplinary effort between biologists and engineers to solve new and old challenges. Ideas abound in the unlikely meeting of people from different disciplines. Let us learn and apply these lessons and, by thus doing, we will honor the treasure chest of knowledge to be found in nature.

2 | HUMANLIKE ROBOTS

Introduction

Over billions of years of evolution, nature has been making trial-and-error experiments within the laws of physics, chemistry, material science, and engineering. The outcome of these experiments is the wealth of fascinating creatures on Earth—including us, the human species.

▲ **Which one is the robot? The Chinese roboticist Zou Renti is on the left, and his clone robot is on the right. This humanlike robot is one of the best attempts so far at making a copy of the appearance of a human, but it is far from representative of the most sophisticated robots around today.**

Humans have always endeavored to mimic and adapt human appearance, capabilities, and intelligence in both art and technology. The field of biomimetics is the latest expression of this, and the desire to engineer machines that display the appearance and behavior of biological humans, and that can perform various functions as efficiently as humans can, represents one of its biggest challenges.

Advances in computer technology, synthetic materials, artificial intelligence, real-time imaging, and speech recognition mean that it is becoming increasingly possible to create lifelike robots that closely resemble humans. Robots that verbally and facially express emotions and respond emotionally are being developed with impressive capabilities and sophistication. Electroactive polymers (EAP), also known as "artificial muscles," are showing promise in enabling the development of biologically inspired mechanisms that once were considered possible only in the realm of science fiction.

This chapter looks at some of the earliest robots that resembled humans and where the inspiration for them came from, then focuses on the state of the art today, the distinctive styles of robot, the challenges ahead, the potential impact of humanlike robots on our society, and the ethical concerns associated with the development and use of robots that are like humans. In order to do this, we will be using two terms to describe two distinctive styles of robot: "humanlike robots" and "humanoids."

HUMANLIKE ROBOTS

These robots are designed to resemble a human as closely as possible. Great efforts are made to copy the exact appearance and behavior of humans. Roboticists making such robots are mostly from Japan, Korea, and China, with a few in the United States. An example of such a robot is shown opposite.

HUMANOIDS

Humanoids are robots with a general human appearance that includes a head, arms, and possibly legs and eyes. They are clearly machines, however, and the head is often featureless or shaped like a helmet. Making humanoid robots is easier for roboticists because they are not required to deal with the complexities of completely humanlike machines.

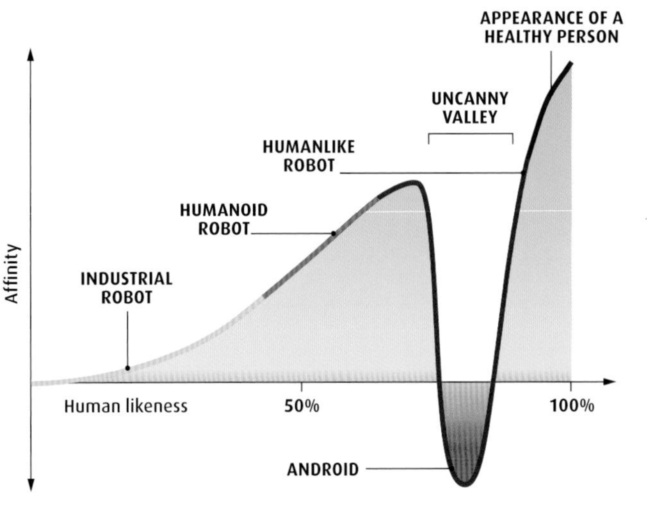

▲ According to roboticist Masahiro Mori, public acceptance of humanlike robots will dip some time after they start to appear human, but it will be regained when the similarity becomes very close.

FEAR AND LOATHING

It is anticipated that improvements in the capability and applicability of humanlike robots will lead to their use as industrial and household appliances. Yet humanlike machines may well invite fear and dislike. The Japanese roboticist Masahiro Mori hypothesizes that, as the degree of similarity between robots and humans increases, people will initially have an enthusiastic response. When this similarity becomes very close, however, it will engender strong rejection and dislike. Only once the similarity reaches a very high level will the response again become positive. In a graphical representation measuring the response to humanlike robots, the negative feeling shows as a marked dip into antipathy, the so-called "valley." The Uncanny Valley theory does have its critics, who suggest that it has never been proven in a credible experiment.

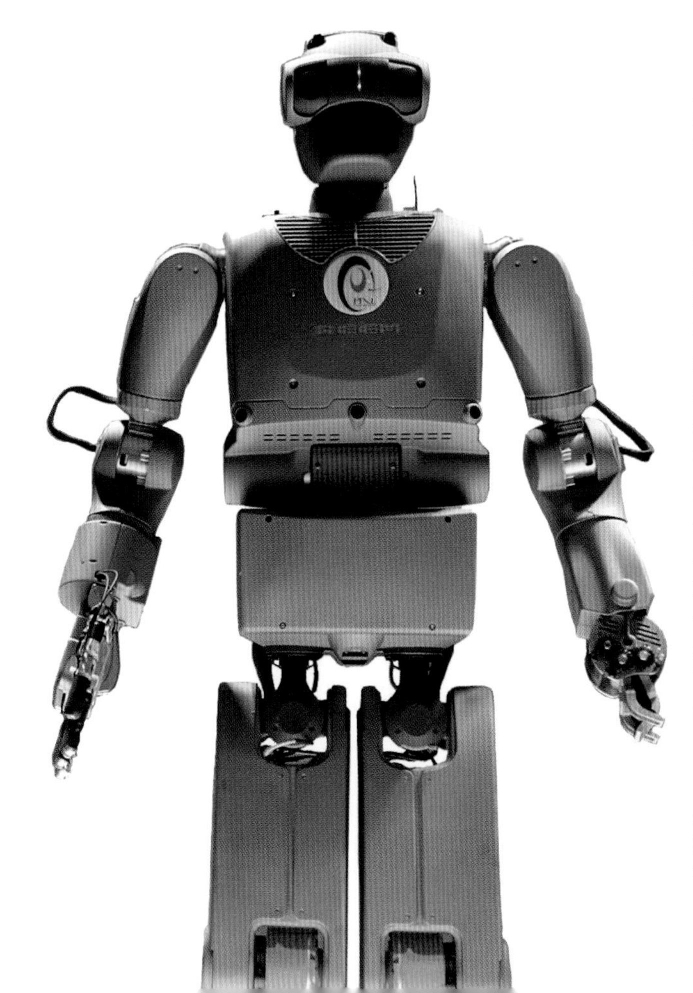

◀ An example of a humanoid robot. While it has a broadly human appearance, with a head, torso, arms, and legs, its features appear more like those of a machine. This robot is called the REEM_A, and it is made by Pal Robotics in Spain.

Historical Review

The Merriam-Webster dictionary defines the word robot as "a machine that looks like a human being and performs various complex acts, such as walking or talking, of a human being." But how did the inspiration for these machines arise, and who decided to call them "robots?"

The first time the word "robot" was used was in 1921, in the play *R.U.R. (Rossum's Universal Robots)*, by the Czech writer Karel Čapek. The word came from the Czech *robota*, which means "compulsory labor," "hard work," or "slavery." The meaning of the word has evolved to become increasingly associated with intelligent mechanisms that have a biologically inspired shape and functions, alongside, in particular, humanlike features.

▲ **In this marble relief, the Greek god of metalsmiths, Hephaestus, is shown with his mechanical helpers, forging the armor of Achilles. This reinforces the idea of robots being associated with hard labor.**

CONCEIVING HUMANLIKE MACHINES

Humanlike machines were envisioned by the ancient Greeks, whose god of metalsmiths, Hephaestus, created his own mechanical helpers in the form of strong, vocal, and intelligent workers. A more recent concept was the golem, a servant made of clay described in Jewish legend from the 16th century, and Mary Shelley's humanlike creation in her 1818 novel *Frankenstein* was a monster made of human body parts and brought to life by the scientist Victor Frankenstein. The common aspect of the golem and Frankenstein stories is that they both describe the construction of a creature to imitate a living human that results in violent and disastrous consequences. The behavior of both these humanlike creations suggests that there is a potential for evil if humanlike forms are created and given freedom to act without control.

DA VINCI DESIGN

Records suggest that, in 1495, Leonardo da Vinci became the first person to make a sketch or plan with a view to producing a humanlike machine. He designed a mechanical knight, known today as "Leonardo's robot." Leonardo's design is related to modern robots only in the

▲ Mary Shelley (1797–1851) published her novel *Frankenstein; or, The Modern Prometheus* when she was 21. It tells the story of a young scientist who creates a human, but who later rejects it as a monster.

▲ A scene from a British television production of *R.U.R. (Rossum's Universal Robots)*. Through this science-fiction play, the term "robot" was introduced into many of the world's languages.

sense that a biologically inspired mechanism was designed to emulate the movement of a human. The machine could sit up, wave its arms, bend its legs, move its head, and open and close its jaws. A mechanism in the chest controlled the arm movements, while an external crank moved the legs.

EARLY ROBOTS

The French engineer and inventor Jacques de Vaucanson is credited with being the first person actually to produce a physical machine that appears and acts like a human. In 1737, he constructed a life-size mechanical musical automaton called the "Flute Player."

Another famous early example of a robot was "The Writer," made by the Swiss clockmaker Pierre Jacquet-Droz in 1772. This humanlike machine acts like a young boy writing at his desk, and is able to write custom text of up to 40 letters in length.

This was an era when many artists with mechanical capability sought to make mechanisms emulating human appearance and movement. Unfortunately, the developers in those days did not have the actuators that we have today, so they used spring-loaded mechanisms. If they had had motors and the control capabilities that we now have at our disposal, they might have been able to create a humanlike robot.

Today's Humanlike Machines

In terms of modern biomimetics, a robot is an electromechanical machine that has biomimetic components and movement characteristics, the ability to manipulate objects, and the capacity to sense its environment. It must also have a degree of intelligence, which is where the key concept of artificial intelligence comes in.

The use of artificial intelligence, to create the robot's "brain," is the key to a robot's capabilities. It allows for the creation of "smart robots," and became possible with the availability of powerful miniature computers. The era of digital computers began in 1946 with the ENIAC computer, the first large-scale general-purpose electronic digital computer. In 1950, the possibility of building machines that could think and learn was raised for the first time by mathematician Alan Turing. Turing did not make such a machine, but through his idea he paved the way for the emergence of real-life robots.

◀ In 1950, British mathematician Alan Turing (1912–54) described the first computer (a Turing machine) in mathematical terms. To this day, the "Turing test" is used to measure the intelligence of machines.

FASTER AND SMARTER ROBOTS

Progress in developing powerful microprocessors with high computation speed, very large memory, wide communication bandwidth, and more effective software tools facilitated the early development of intelligent robots. Today, the rapid processing of computer codes and effective control algorithms are allowing the development of increasingly sophisticated humanlike systems and robots. Operating a robot in natural environments, however, is a complex task. There are so many unpredictable obstacles that it is virtually impossible to preprogram a robot for every foreseeable circumstance. The robot needs to be able to deal with complex situations on its own, adapt, and learn from its experience. In order to develop such sophisticated capabilities, artificial intelligence (AI) researchers are using methodologies that are inspired and guided by nature.

INSTILLING COMMON SENSE

Incorporating intelligent operation into a humanlike robot is one of the most important aspects of making it lifelike. This is also guided by nature because no creature can survive if it does not have some level of intelligence in dealing with its surrounding environment and the hazards involved. Imitating human functions can be as simple as moving various parts, but to be useful it is essential to have an artificial equivalent of human intelligence.

LIFELIKE REQUIREMENTS

The rapid strides being made in constructing lifelike robots are made possible by a variety of technologies in different disciplines. These include:

- state-of-the-art microprocessors
- effective autonomous operation algorithms (sets of rules)
- humanlike materials, such as Frubber skin
- movement simulators
- sensors to imitate the senses

Even the related software used in robots increasingly resembles the organization and functionality of the human central nervous system, and it helps robots to perceive, interpret, respond, and adapt to their environment.

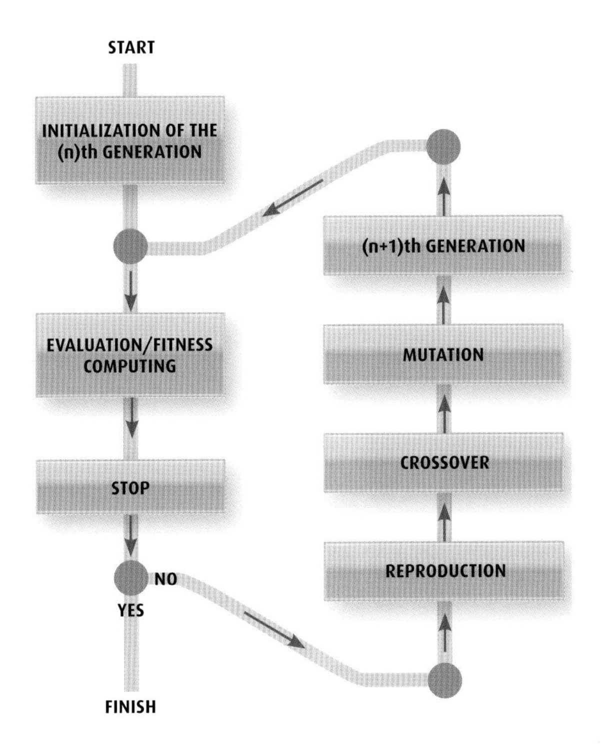

▲ This example of a biologically inspired genetic algorithm shows how the evolutionary process of natural selection can be applied to pairs of mathematical combinations, to weed out unfit members and produce new, more promising pairings for improved future generations.

THE INTELLIGENT ROBOT

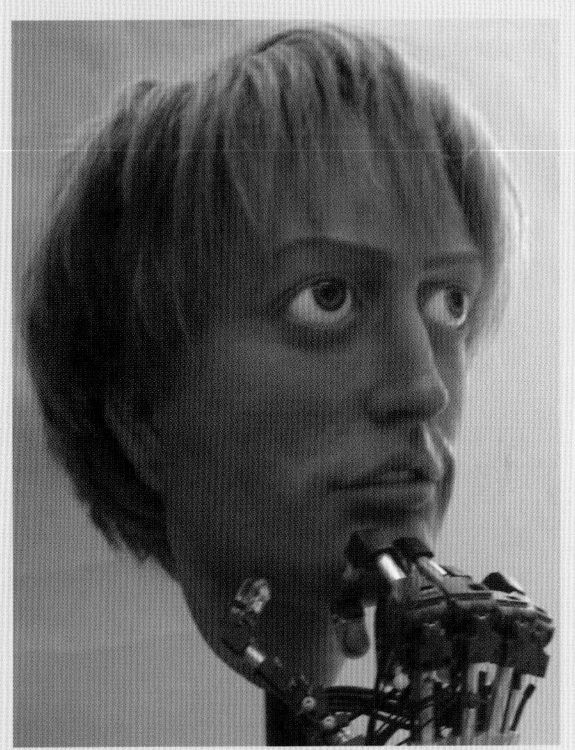

Instilling intelligence into a humanlike robot requires sophisticated artificial intelligence algorithms. The head of this robot was made by David Hanson and the hand by Graham Whiteley, and they both are located in the Jet Propulsion Laboratory (JPL) at California Institute of Technology.

Genetic algorithms are an example of such biologically inspired systems. They simulate natural selection—survival of the fittest—where a population of individual mathematical combinations evolves over time. Pairs of combinations interact and produce "offspring" combinations. If they work well, they are retained, while unfit members are eliminated. At the end of the evolution process, the population usually consists of fairly good solutions—but there is no assurance that the optimal solution has been found.

The Need for Humanlike Robots

Robots are entering our lives in education, healthcare, entertainment, domestic assistance, and military applications. Currently, entertainment robots are the most prevalent, with humanlike robotic toys widely available. One example is the Zeno interactive learning companion, a kind of synthetic pal.

The movie industry has begun to collaborate with scientists to make robotic characters appear more realistic and ensure that they move more like people. Increasingly, robotics researchers are collaborating with artists to make their robots appear more expressive, believable, and heartwarming. The more we develop and interact with humanlike robots, the more sophisticated and lifelike they are becoming. What's more, in the process, we are gaining a greater understanding of ourselves—scientifically, socially, and ethically.

▶ Zeno is a robotic friend and learning companion for children. He can link up to a computer, walk, talk, and recognize who you are. His soft face also shows emotions.

WHERE PEOPLE CANNOT GO

In industry, robot applications include planetary or deep ocean exploration. Robots can operate in areas that humans cannot reach, such as places with toxic gases, radioactivity, dangerous chemicals, bad odors, biological hazards, or extreme temperatures. Within these areas, robots could be used for clearing hazardous waste, explosives disposal, and search and rescue operations.

To be capable of operating autonomously in these environments, a robot needs to perceive its environment, make decisions, and perform complex tasks independently, just as a person would. Advances in this area have been enormous in recent years.

BEING HUMAN

The world we have built around us has been made to fit our body size, shape, and capabilities. This includes our home, workplace, and facilities, the tools we use, and the height at which we keep various items. Therefore, the robots that are being made to help us would best operate if they matched our shape, average size, and capability. Such configuration will allow robots to reach handles to open doors, listen to us at eye level, climb stairs, sit on our chairs, drive our vehicles, and perform many other support

▲ This multilegged robot is designed to traverse complex terrain. It was developed in NASA's Jet Propulsion Laboratory.

tasks for us. Also, since we respond intuitively to body language and gestures, it is highly desirable for robots to use facial and body expressions.

GETTING AROUND

Constructing a robot that looks and behaves like a human is only one level of the complexity of this challenge. Making robots that can physically function within and outside our homes will involve the need to have them successfully navigate complex terrains. This involves dealing with static objects such as stairs and furniture, and responding to dynamic ones such as people, pets, or automobiles. Tasks such as walking in a crowded street, crossing a street with traffic while obeying pedestrian laws, or walking in a complex terrain with unpaved roads require determining an available path that is both safe and within the robot's capability. Technology to achieve this is the subject of research at many robotics laboratories worldwide.

WALKING LIKE A HUMAN

This humanoid robot, called Chroino, is made of carbon and plastic, which gives it a light but strong structure known as a "monocoque frame." It was made by developer Tomotaka Takahashi at Kyoto University's Robo Garage, in Japan. The name "Chroino" comes from the words meaning "to chronicle" and "black," the second of which is pronounced *kuroi* in Japanese. Below are two of the robot's key features:

❶ Incorporates new "SHIN-Walk" technology, enabling it to walk in a more natural way, almost as smoothly as a human does.

❷ Lightweight: it weighs just 2⅓ pounds (1.05 kg).

Making a Humanlike Robot

Another key challenge in creating machines that mimic humans is to develop robots that have the ability to interact with people and to communicate emotionally. By providing opportunities for interactive exchanges, there is the potential for robots to stimulate communication skills in humans.

Children today are spending more time with computers than with peers of their age or with people in general; they are growing up with less developed social skills and a poorer understanding of body language cues than those that were taken for granted in previous generations. This growing concern may be addressed by incorporating humanlike robots into education, therapy, or games, while providing realistic simulation under controlled conditions. An example of one such robot, Zeno, is shown on page 52. Robots are already being used in studies of treatments for children with autism and reported results show great promise.

BATTLEFIELD ROBOTS

With the increase in our ability to make humanlike robots more lifelike, however, there is a growing concern that they will be used for improper tasks. Inevitably, humanlike robots will be designed for military applications. Currently, under a U.S. Defense Advanced Research Projects Agency (DARPA) program, the focus is on the development of a robotic hand that can be controlled in various ways, but the capability could be expanded to other body parts.

An example of a robotic arm is shown opposite. Such robotic arms are being developed for use by the military. Using robots against enemy humans raises ethical and philosophical issues, and practical dangers. These would need to be resolved at the same time as the technology is enhanced. Guidelines, such as science-fiction writer Isaac Asimov's "Three Laws of Robotics" (see opposite), have long been suggested. But there are many less controversial uses of this technology, such as robotic arms and hands for amputees, which require less stringent controls.

◀ **Humans can pick up and manipulate delicate objects between their fingers, and robots need to be able to mimic this capability in order to work well.**

▲ This robot, called Twendy-One, was designed at Waseda University in Tokyo. Here it shows its remarkable ability to pick up and manipulate a delicate drinking straw between its fingers without crushing it.

HUMAN ABILITIES

The complex task of making humanlike robots involves not only copying the appearance of humans, but also replicating human capabilities such as the ability to communicate emotions and thoughts. This task involves many science and engineering disciplines, including mechanical and electrical engineering, materials science, computer science, artificial intelligence, and control. It requires materials that are resilient, lightweight, and multifunctional. Robots need to be able to walk while avoiding obstacles and maintaining high stability. A lightweight, mobile, and durable energy source is essential. Humanlike robots also need sensors to replicate the human senses of sight, hearing, taste, smell, and touch, including the sensing of pressure and temperature. They need to be able to interpret these sensors' measurements so that they can perceive their environment and its related hazards. The integration of these required capabilities will make the robot a "smart" machine that looks and acts like a human.

ASIMOV'S LAWS OF ROBOTICS

In his famous "Three Laws of Robotics," to which he later added his Zeroth Law, the well-known science-fiction writer Isaac Asimov suggested guides for human and robot relations. In these proposed laws, he suggested that robots would function in the role of servants and should not be allowed to cause harm or injury to humans.

1. A robot may not injure a human being or, through inaction, allow a human being to come to harm.
2. A robot must obey orders given to it by human beings, except where such orders would conflict with the First Law.
3. A robot must protect its own existence as long as such protection does not conflict with the First or Second Law.

Zeroth Law A robot must not act merely in the interests of individual humans, but those of all humanity.

Robot Parts

A robot needs to have certain features and abilities in order to appear humanlike. For example, in order to communicate in a humanlike manner, the robot needs a voice with synchronized lip movements that are coordinated with body gestures, and the skin needs to be sufficiently elastic to allow facial expressions.

Of utmost importance, a robot needs to operate safely in its interaction with humans and surrounding objects. Here are some of the features necessary to make it appear humanlike.

BRAIN

The "brain" is made up of multiple microprocessors, a miniature computer that is used to control the operation of the robot, including mobility, image processing, verbal communication, body language, and facial expressions, as well as many other tasks that the robot performs. The capability of the available microprocessors is continually increasing at an enormous rate, where the speed of processing and the volume of data have been growing many times each year, while cost and size have been steadily decreasing. A robot's "brain" does not have to be mounted inside the head, but placement here can minimize unnecessary wiring to some of the critical sensors.

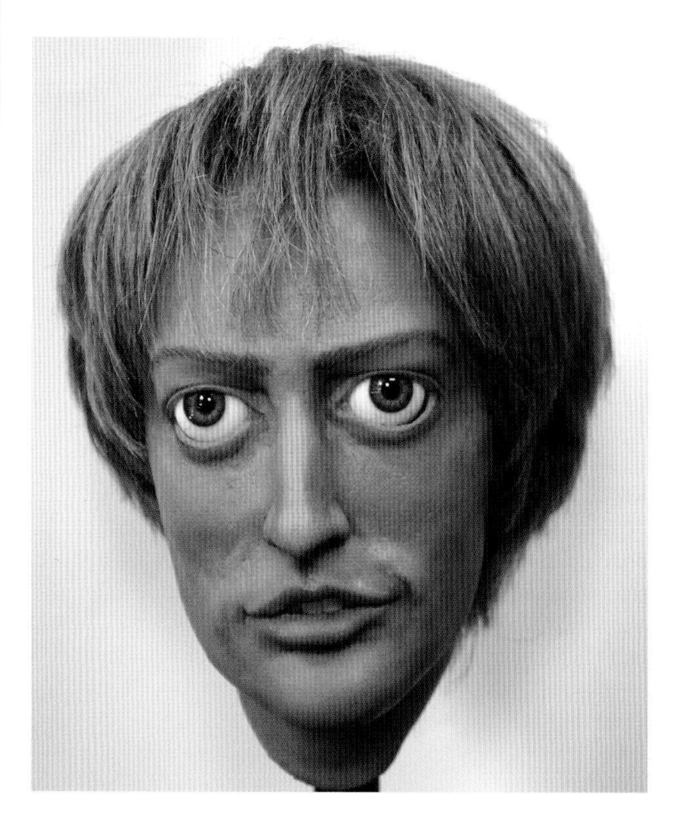

▶ This robotic head, which was made by David Hanson of Hanson Robotics, is covered with Frubber skin, to make it more humanlike. The head is capable of a range of facial expressions and has been used for testing artificial muscles.

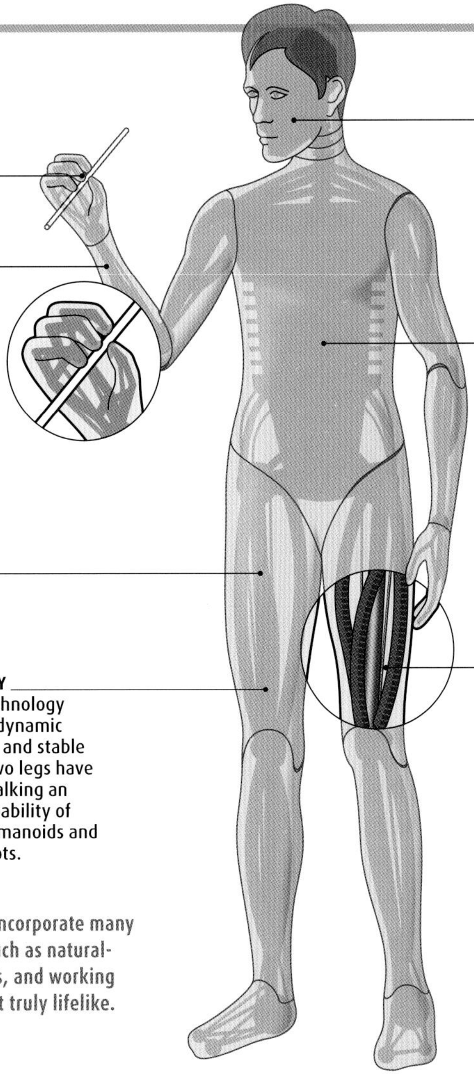

ARMS, HANDS, AND LEGS
To mimic natural human movement, sensors are mounted on the arms, hands, and legs, including pressure sensors that determine grip pressure, and touch sensors to interpret tactile impressions. Sensors are also used to trigger responses when the robot is exposed to unsafe conditions.

FACE
The developer needs to be able to generate lifelike expressions that include wrinkles and folds, which are dormant in the nonexpressive position of the face.

SKIN
To appear humanlike, the robot needs to be covered with an artificial skin that looks and feels like the skin of a living person. This requires the skin to be highly elastic, enabling facial expressions without residual deformation.

BIPED STABILITY
Advances in technology in the areas of dynamic balance control and stable operation on two legs have made stable walking an established capability of many biped humanoids and humanlike robots.

ARTIFICIAL MUSCLES
Actuators emulate muscles and are responsible for robots' mobility and the movement of their appendages and other parts and mechanisms. The types of actuator that are generally used for these include electric, pneumatic, hydraulic, piezoelectric, shape-memory alloys, and ultrasonic devices.

▶ A humanlike robot has to incorporate many components and materials, such as natural-looking skin, pressure sensors, and working "muscles," in order to make it truly lifelike.

THE SENSES

The face can include an artificial nose and tongue to provide the humanlike robot with information about smell and taste. Vision is provided via video cameras, allowing the robot to "see" its environment and location, and to receive communication cues from facial expressions—these support its ability to act sociably. Sound is sensed for content and direction, with speech recognition supporting natural verbal communication. Pressure sensors allow the robot to "touch" and "feel" things.

ARMS, HANDS, AND LEGS

The arms, hands, and legs of humanlike robots have the same basic functions as they do in humans. These may seem easy to replicate, but they are hard to control in robots.

ACTUATORS AND ARTIFICIAL MUSCLES

Electric motors are used to produce the movements of humanlike robots, but they behave differently from our natural muscles and have a totally different operating mechanism to our compliant and linear natural muscles.

Key Technologies

Several key technologies are essential for constructing a humanlike robot that is convincingly lifelike, including certain materials, actuators, sensors, smart control, mechanisms of mobility, manipulation of the hands and legs, hearing, verbal communication, seeing and image interpretation.

The humanlike robot needs to have the ability to determine obstacles and risks, as well as to avoid and overcome them. There must be effective control and artificial intelligence algorithms in place to allow it to operate in humanlike form, as well as to interact with its environment and with humans. In order to achieve its biomimetic purpose, it needs to have effective body parts and related functions as similar as possible to those of a real human.

ELECTROACTIVE POLYMERS

The actuators that are closest to emulating natural muscles are the electroactive polymers (EAP) that have emerged in recent years, which have gained the name "artificial muscles." Many of the EAP materials that are known today emerged in the 1990s, but they are all still weak in terms of their ability to perform significant mechanical tasks such as lifting heavy objects. Recognizing the need for international cooperation, the author of this chapter organized the first annual international EAP Actuators and Devices (EAPAD) Conference in March 1999. At the opening of the first conference, the author posed a challenge to scientists and engineers worldwide: to develop a robotic arm actuated by artificial muscles that was capable of winning an arm-wrestling match against a human opponent (see left).

◀ **Robotic technology has come a long way since this arm-wrestling competition between a young high-school student and a robotic arm in 2005. The girl won.**

◀ Robotic arms vary considerably in appearance and abilities. This example was photographed by the author at the DARPATech symposium held in Anaheim, California, in 2007.

WRESTLING CHALLENGE

The first arm-wrestling competition between a robotic arm and a human—a 17-year-old female high-school student—was held on March 7, 2005. Three robotic arms participated in the contest, and the girl beat all of them. The second Artificial Muscles Arm-Wrestling Contest was held on February 27, 2006. Rather than the contest consisting of wrestling with a human opponent, a measuring fixture was used to test the EAP-actuated arms for speed and pulling force. To establish a baseline for performance comparison, the capability of the student from the 2005 competition was measured first. The 2006 results were two orders of magnitude lower than that of the student. In a future conference, in the light of advances in developing such arms, a professional arm wrestler will be invited to take part in the next human vs. machine wrestling match.

ARTIFICIAL SKIN

One of the most notable artificial skins that has been developed is Frubber, created by David Hanson. An example of a robotic head made with this elastic polymer is shown on page 56. This rubbery material requires minimal force and power to produce natural-looking large deformations. This is serving as a platform for engineers worldwide who are developing artificial muscles and need to test their developed actuators.

PROSTHETICS

One benefit of the success in making humanlike hands, arms, and legs for robots has been the emergence of highly effective and lifelike prosthetics. Another area of development is in walking chairs as a substitute for wheelchairs, enabling users to traverse areas that are not flat. An example of such a chair was demonstrated at the 2007 NextFest Exhibition of *Wired* Magazine. It carried a person while it was walking and climbing stairs. This ground-breaking invention will help to give disabled people greater mobility over uneven ground.

▶ The latest research in robotics may lead to prosthetic limbs that can not only be controlled by a person's nervous system, but also provide feedback so that the person can feel what the arm or hand, or foot or leg, is touching.

Artificial Intelligence

Artificial intelligence (AI) is a branch of computer science that studies the computational requirements for such tasks as perception, reasoning, and learning, to allow the development of systems that perform these capabilities. A robot needs to be given these capabilities in order to be truly humanlike.

The objectives in the AI field are to advance the understanding of human cognition, to understand the requirements for intelligence in general, and to develop intelligent devices, autonomous agents, and systems that cooperate with humans to enhance their abilities. AI researchers are using models that have been inspired by the computational capability of the brain and explaining them in terms of higher-level psychological constructs, such as plans and goals.

Progress in the field of AI has enabled better scientific understanding of the mechanisms underlying thought and intelligent behavior, and their embodiment in robots. The field of AI is providing important tools for humanlike robots, including knowledge capture, representation and reasoning, reasoning under uncertainty, planning, vision, face and feature tracking, language processing, mapping and navigation, natural language processing, and machine learning. AI-based algorithms are used where case-based and fuzzy reasoning are combined to allow for automatic and autonomous operation. Even though AI has led to enormous successes in making smart computer-controlled systems, the capability is still far from resembling human intelligence.

◀ **This child robot has soft silicone skin and is learning to think like a baby. It evaluates the facial expressions of its "mother" and is slowly developing social skills.**

ESSENTIAL SKILLS FOR A ROBOT

When programming a robot, the following steps are needed, in which AI plays an important role:

1. Sensing the surrounding environment.
2. Modeling the environment using inputs from computer vision, hearing, and other sensors.
3. Planning a robot action accounting for obstacles and dangers along its potential paths.
4. Taking appropriate actions to achieve the goals of the robot's operation.

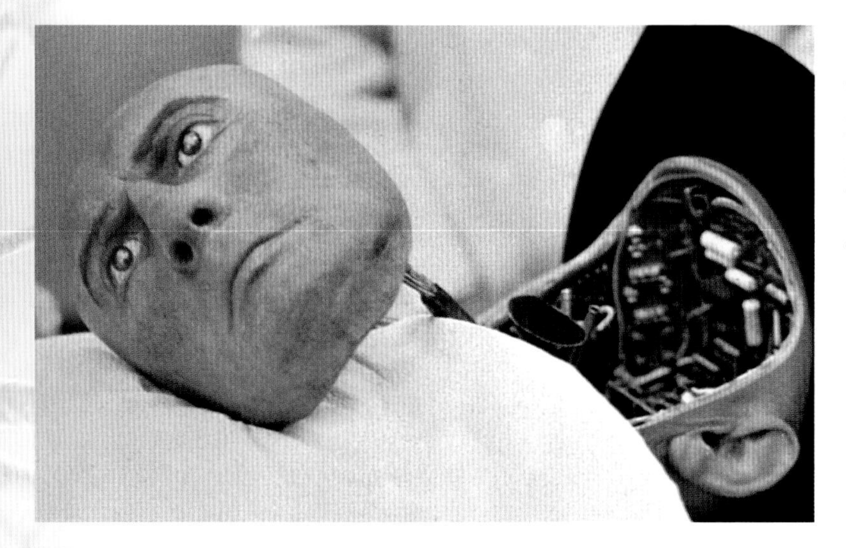

◀ A still from the movie *Westworld* (1973). The action focuses on a theme park where humans pay to act out their gunfighting fantasies. The robots are programmed to lose, but things go wrong when the wiring of the main robot, known as "The Gunslinger," breaks down and the machine is freed to use its intelligence to win.

ROBOTIC SELF-PRESERVATION

Some of the latest humanlike robots demonstrate a very human trait: the ability to improve themselves even after their production, since they can self-learn and obtain periodic updates. The sophistication of these robots includes fully autonomous operation and self-diagnostics. In the future, they may be designed to travel on their own to a selected maintenance facility for periodic checkups and repairs. In case of damage, future robots may be made of biomimetic materials that are capable of self-healing.

A LONG WAY TO GO

The capability of current humanlike robots is still far from matching that of humans or the portrayals offered by science-fiction books and movies. Progress has often moved more slowly than experts have predicted. For example, in the 1950s, AI experts forecast that a computer would beat the world chess champion by 1968, but it took many years longer for that prophecy to be fulfilled. Yet there is no doubt that artificial intelligence is already all around us. Every mobile phone call we make, and every email we send, is already being routed using AI systems. And writers' imaginatively conceived creations continue to inspire and guide innovation in the development of humanlike robots, while also alerting us to the dangers and negative possibilities along the way.

▶ The expressiveness of this female android's facial features and gestures conveys a sense of her intelligence and responsiveness to the world at large.

Practical Applications

Progress in voice synthesis, detection, and recognition continues to facilitate interaction between humanlike robots and humans. Integrated technology is enabling robots to communicate verbally, expressing emotions while making eye contact and facial expressions, as well as responding to emotional and verbal cues.

New technology is advancing emotional adaptation in robots as they interact with people in ways that are familiar to humans, without the need for training. Robots can mimic humans by nodding when listening to someone speaking to them, periodically blinking, and looking a speaker in the eye for brief sessions. Conversations with robots are currently limited to a vocabulary of about 1,000 words, and content and topics are restricted, but a significant amount of research is dedicated to the development of biomimetic machine capability in understanding human conversation.

◀ Robotic arms are ideally suited for carrying out spot welding and arc welding in assembly lines for truck or other vehicle bodies.

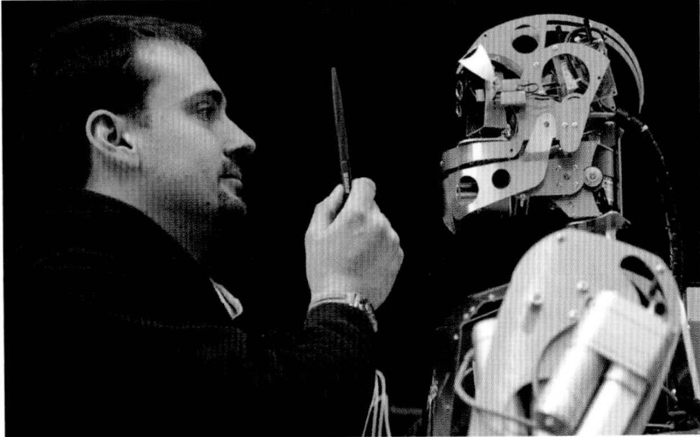

▲ A technician adjusts a humanoid robot at the 2009 Hannover Messe, an annual industrial trade fair, in Germany; a total of 6,150 companies from 61 nations took part.

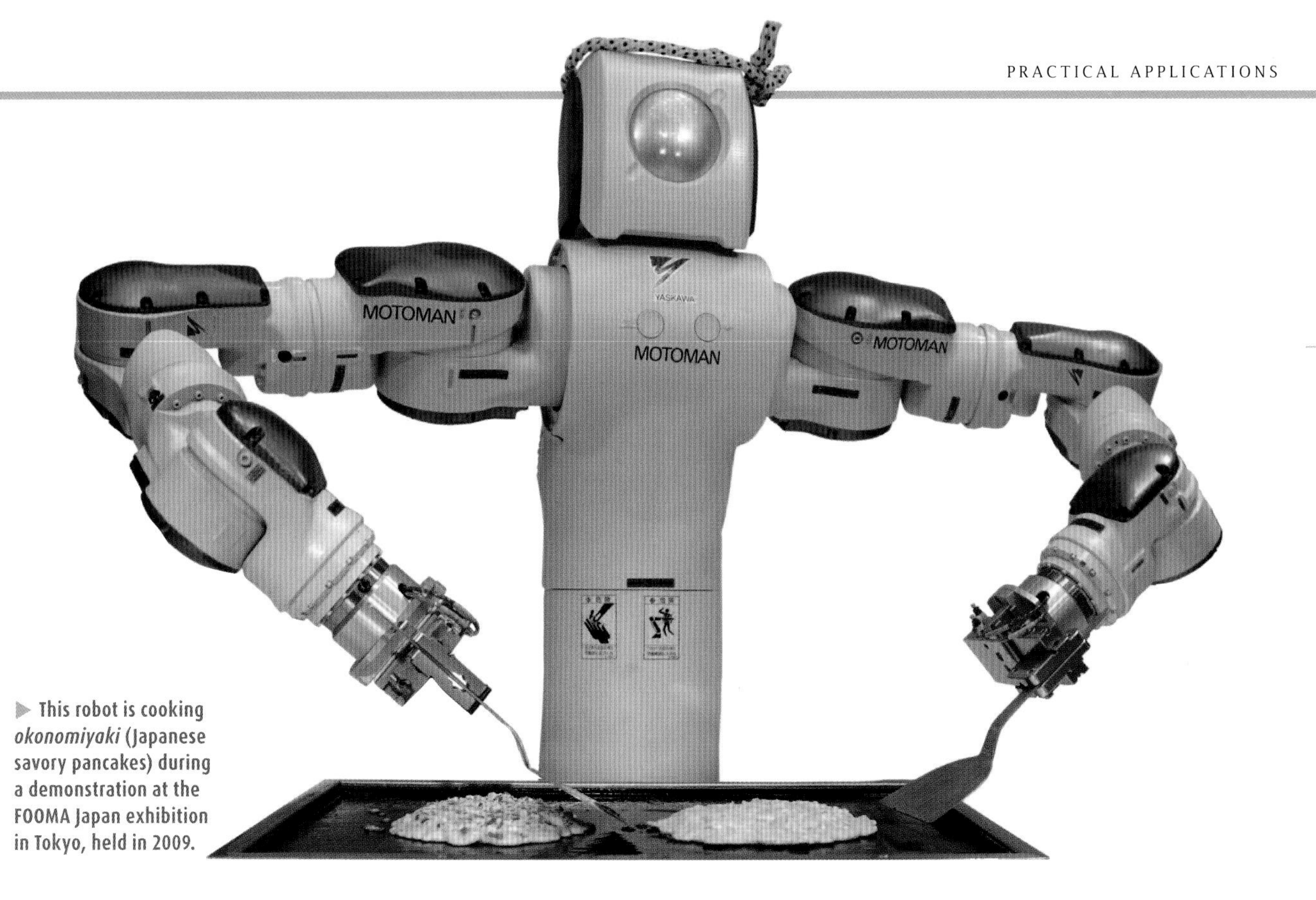

▶ This robot is cooking *okonomiyaki* (Japanese savory pancakes) during a demonstration at the FOOMA Japan exhibition in Tokyo, held in 2009.

TELEPRESENCE

Even within their current limited sphere of intelligent operation, there are various practical applications for humanlike robots, such as telepresence, which is the ability to operate a robot remotely in a way that enables the operator to act as if he or she is present at the robot's location. One example is the Robonaut (robotic astronaut) at NASA's Johnson Space Center (JSC), in Houston, Texas. This humanlike robot can mirror the physical movements of the upper section of the human body. It is operated remotely outside the space shuttle or space station by a control user on Earth or inside the space vehicle; its uses now include potential military applications.

MEDICAL ROBOTS

The increasing capabilities of humanlike robots offer promising medical applications. For example, robots are being developed in Japan and the United States to assist recovering patients, the elderly, and other people who need physical or emotional support. Robotic surgery is increasingly becoming part of the many capabilities that are available to surgeons today, and the results are encouraging medical specialists to use it more. As assistive devices, humanlike robots are used to remind patients to take their medication and to bring it to them; they serve as a tool for entertainment, as well as a monitoring system, sending, in emergencies, live images to the staff at a central control room.

CURRENT LIMITS

Even though developments in the technology of humanlike robots are progressing rapidly, there are still issues that limit their widespread use, including limited functionality, relatively short battery charge, and high cost. Once they reach mass production levels and become affordable, however, we can expect to see them become common household helpers, performing valuable human-related services.

Can Robots Evolve to Become Human?

With artificial intelligence, humanlike robots are capable of facial recognition and have personalized behavior that varies, even between duplicates of particular robots. Some robots can walk or dance in a similar way to a human. There are, however, many basic human tasks that are beyond robots.

Human tasks that are beyond robots include conducting a comprehensive conversation with a human on a broad range of subjects, walking fast in a crowd without hitting anyone, and operating over an extended period of time. The process of meeting these challenges is expected to be evolutionary—a suitably humanlike trait—with a series of successes that will lead to more realistic performance.

COST IMPLICATIONS

An important factor in the development of humanlike robots is low-cost production, and this requires the availability of standard hardware components and software platforms that are interchangeable and compatible. This standardization can follow the same pattern of development as personal computers, where today parts and software are made to be independent of the specific computer manufacturer. This will allow scientists and engineers to avoid the need for expertise that is too broad, and thus focus on making improvements in their area of specialization. In addition, hardware and software will need to provide humanlike robots with higher speeds of response to changes in their environment. Robot hardware will need to be significantly reduced in weight and equipped with numerous miniature lightweight actuators and sensors with distributed processing capabilities. Effective actuators with high power density, as well as high operational and reaction speeds, are required. There is also a need for advances in automated design and prototyping, from miniature actuators that allow the making of miniature motors to control fine movements for facial expression or movement of fingers in a baby-size robot, to microscale sensors and drive electronics.

The ability of future robots to conduct a comprehensive conversation will require the capability to recognize and "understand" more words than they can today, with significantly higher accuracy of interpreting text and verbal communication. The use of networked wireless robots will harness the power of personal computers to handle complex tasks, including image and speech recognition, navigation, and collision avoidance.

▶ Robot armies may initially be operated remotely, and in the future may become totally autonomous. Robots do not suffer from human deficiencies such as post-traumatic stress, or emotions such as fear.

◀ Even with all the recent advances in technology, one of the basic human skills that still defeats the current capabilities of robots is the ability to walk quickly through a crowd without colliding with anyone.

RIGHTS AND WRONGS

Making a humanlike robot self-aware of the consequences of its actions and having it operate with rules of "right" and "wrong" would give it a much more humanlike character, but some claim that this is impossible. The problem centers on the robot having to adjust its behavior to take into account subjective uncertainties; this involves a sense of proportionality that is beyond literal interpretation of situations. One fascinating ethical concern is the possibility of robot nonobedience and unacceptable behavior, despite the concept of robots being developed to operate with humans in master–slave relationships.

TEAM EFFORT

Making a humanlike robot is multidisciplinary, requiring expertise in such fields as electromechanical engineering, computational and material sciences, neuroscience, and biomechanics. Advances in AI, effective actuators, artificial vision, speech synthesizers and recognition, mobility, control, and many other fields are aiding significantly in making robots act like humans. And these developments spin off into many other fields and areas of application.

RELATIONS WITH HUMANS

Increasingly, robots in various shapes, including rovers and drones, are being used for military applications, and it is inevitable that humanlike robots will be used for such purposes, too. If they are developed with artificial cognition and surpass human levels of intelligence, however, one may wonder whether they may potentially turn against their human creators.

As for creating an army, genetic cloning will not lead to an instant and exact duplicate of a human. The produced person needs to grow biologically and naturally. Humanlike robots, though, may be produced at the speed of manufacturing processes. As rapid prototyping fabrication of humanlike figures becomes more effective, our neighborhoods may one day be filled with robots.

3 | UNDERWATER BIOACOUSTICS

Introduction

How can you tell whether a watermelon will be sweet without tasting it? Each watermelon has a scar on the opposite end from where its stem was; if the scar is small, the watermelon is likely to be sweet. Tapping on the watermelon and listening to the sound it makes is another very effective way of "tasting."

A bad watermelon will make a dull sound with many different frequency components because the melon has numerous "cracks" inside that prevent sound from propagating symmetrically. In contrast, a sweet watermelon sounds like a bell, with a relatively pure tone and few frequencies, because the sounds are not intercepted by cracks and hence propagate symmetrically. And so, without cutting and tasting the watermelon, you can "see" what it is like inside just by listening to the echo.

▲ A sharp tap on a watermelon produces an echo that enables us to "see" what the watermelon is like inside by listening to the sound.

USING AUDIBLE CLUES

When riding a bicycle on a busy road, you will move to one side to avoid an automobile approaching from behind, even though you do not see it. This is because the vehicle's sound alerts you in advance to its approach. Sounds such as these provide helpful signals to us in everyday life, and we learn by experience to respond to them automatically.

MATCHING SOUNDS WITH IMAGES

Listening and imaging are not just a question of physics; they are also based on learning and experience. You have a huge database in your brain to match sounds with specific targets. So the echo from a watermelon tells you if it is sweet, while the noise of an approaching automobile keeps you safe, even when you do not see the vehicle behind you. Sound is indeed a useful tool for obtaining information about invisible targets in our daily life.

REINSTATING IMPORTANT SOUNDS

Advances in technology often lead to quieter performance in objects around us, but this is not always helpful. For example, hybrid or electric automobiles are becoming more common. These near-silent vehicles may lead to more accidents because people will not hear them approaching.

Requiring that quiet vehicles generate some sort of artificial engine noise may be one way of resolving this problem, working in much the same way as the artificial clicking sound that some automobiles now make when their indicators are blinking. Indicators were originally controlled by a mechanical relay that made an actual clicking sound, but today the electronic circuit that controls the indicator lights makes no sound at all. Automobile manufacturers have added the artificial "click, click, click" sound as a safety feedback feature to let the driver know when the signals are activated.

▲ Sounds act as signals to alert us to events around us. The sound of breaking glass, for example, brings to mind an immediate warning image of the presence of dangerous shards of glass.

▶ Being able to hear the sound of an approaching vehicle and interpret its size, distance away, and speed before it is visible is crucial in helping a pedestrian or cyclist to stay safe on an open road.

DRAWBACKS OF USING SOUNDS TO "SEE" IMAGES

Sounds are very helpful, but our automatic responses to them do have their drawbacks. For example, an English cyclist in London instinctively moves to the left to avoid an automobile coming up from behind because vehicles drive on the left-hand side in England. This automatic response could have fatal consequences for the same cyclist traveling on a busy road in New York because traffic moves on the right-hand side of the road in North America. The cyclist could swerve in front of the vehicle without conscious thought, and cause an accident. Clearly, then, sounds are very useful signals, but they must be used with caution.

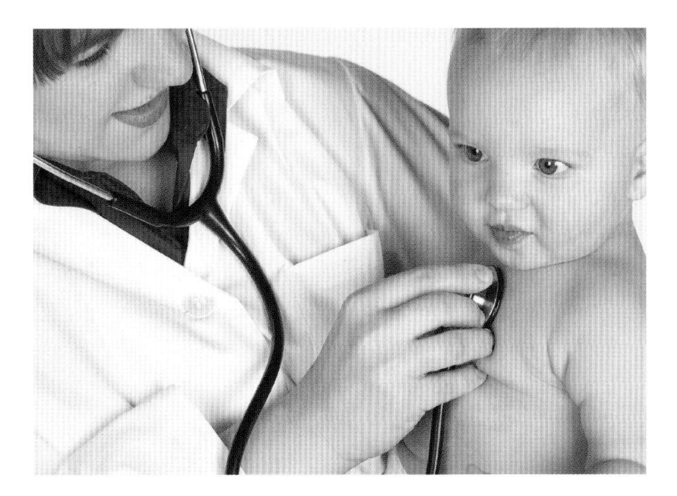

▶ By listening through a stethoscope, a medical doctor is able to "see" a mental image of the interior of a baby's chest cavity and make a diagnosis.

Making Waves in the Water

Long before human beings used sounds to "see" invisible targets, dolphins and porpoises depended on ultrasonic pulse sounds to catch prey in murky waters or darkness. Engineers are trying to understand how these work so that they can develop new echo sounders that use sound to detect shapes and locations of objects.

In the muddy yellow waters of the Yangtze River in China, nothing can be seen farther than 20 inches (50 cm) away. To cope with this lack of visibility, the Chinese river dolphin (*Lipotes vexillifer*) or baiji, which is considered to be extinct, used ultrasonic sounds to sense prey and obstacles. The baiji and other toothed whales produce high-frequency sounds and listen for echoes to spot their prey. It was natural for toothed whales such as dolphins and porpoises to evolve this biosonar ability because they pursue each prey and catch it, one at a time, in the mouth. With biosonar, they can determine distance, size, material composition, and even shape and internal structure of underwater objects, and distinguish between them.

SOUNDING OUT THE SURROUNDINGS

Humans have expended a lot of effort on developing ways to identify targets under the water's surface. Underwater poachers or covert operators with diving gear are very difficult to detect. A sensing system that could detect an unauthorized diver in restricted fishing areas or near an atomic power plant would be a valuable security measure. It would also help fishermen to locate fish and ascertain what species they are, before letting down their nets. The price of fish greatly depends on the species and size. Unfortunately, radio waves and light are not very effective for this because they do not travel long distances in water.

◀ **The wonderful sonar ability of the dolphin could help scientists develop an underwater sensing system that could detect poaching divers and other unwelcome intruders.**

▲ The Chinese river dolphin had very small eyeballs with atrophied crystalline lenses and was unable to see visual images clearly, so it used ultrasonic sounds to locate prey and obstacles.

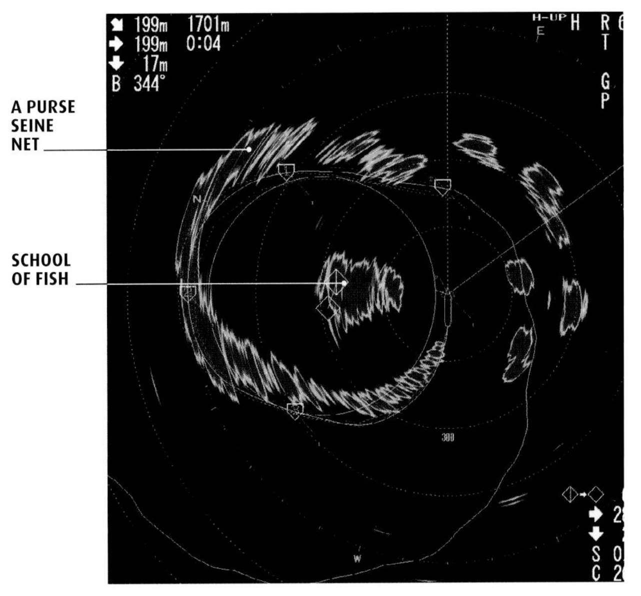

▶ A scanning sonar image of a school of fish, which is about to be caught by a purse seine net. Sound is an effective tool for helping to visualize underwater targets, which is why dolphins and humans use it.

With the naked eye we can see an aircraft in the sky from several miles away, but beneath the surface of the ocean we cannot see a submarine only 100 feet (30 m) away. Light is not effective for long-distance underwater sensing, but sounds are. Sound travels almost five times faster in water than it does in air and is less attenuated while traveling. This is why dolphins and porpoises use acoustic sensory systems for quick and long-distance searching in the water. Their most important target is fish; their sonar probably evolved to find and classify target fish suitable to eat. We can learn from dolphins' sophisticated biosonar systems as we develop echo sounders for fisheries to locate and identify shoals of fish.

Listening to the "Noisy World"

In 1956, the French explorer and scientist Jacques-Yves Cousteau and film director Louis Malle released a documentary film called The Silent World. *The title, however, turned out to describe the complete opposite of what their amazing film showed. The "silent" ocean is in fact a very noisy place to be.*

The real underwater world is noisy. Many fish and shellfish produce sounds, and snapping shrimp make the most noise of all, snapping their pincers and generating high-intensity pulse sounds with broad frequency components at all times. The Japanese call this "tempura" noise because it sounds like the noise made by deep-frying. Ocean waves also create bubbles that generate broadband sounds. Thermal noise caused by molecular motion is another dominant noise source in the ultrasonic range.

▶ **It is extremely difficult for humans to speak intelligibly in an aquatic environment, even with underwater microphones, because the boundary between air and water inhibits transmission of sound. That is why scuba divers use hand signals to communicate with one another underwater.**

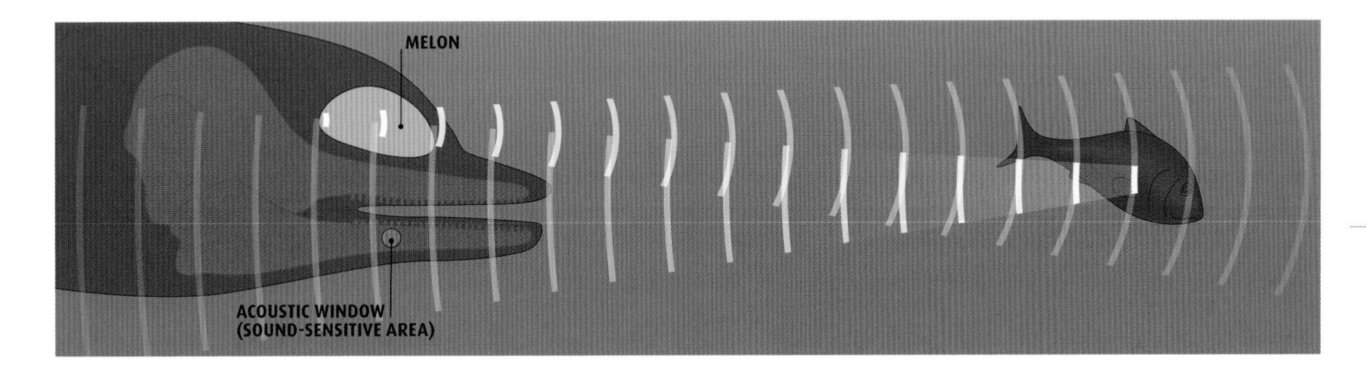

▲ A dolphin focuses clicking sounds through the "melon" organ ahead of its skull. The sound travels forward through the water to the target fish; echoes that bounce back are received through the dolphin's jaw and into its inner ear system. We still do not know whether the dolphin recognizes the visual image or translates from acoustical cues in the echo.

HEAR FOR YOURSELF

If you put your head under a couple feet of water, you will not be able to hear the voices of people outside the water because the water surface reflects all the sound. In the same manner, you cannot hear people talking on a boat when you dive into Cousteau's "silent" world wearing scuba gear. All you will hear is the sound of bubbles as you breathe.

The human ear is not designed to receive underwater sounds. The amplitude of the sound oscillation in water is generally too small to vibrate your eardrum. In addition, air remains in the ear canal and prevents sound energy from reaching the inner ear. You can still listen to tempura noise in the water, but it will be difficult to identify the source direction because the sound pathway to your inner ear is different. Most of the underwater sound you hear comes by bone conduction through your skull or your jawbone; this is much more efficient because it requires no water-to-air conduction.

Since the difference in density between water and flesh or bone is much smaller than the difference between water and air, sounds are not reflected as much at these boundaries. Speaking underwater is also difficult for humans. Words spoken underwater are hard to understand. They do not project into the water well because the boundary between air and water acts a barrier to the transmission of sound.

SOUNDING THROUGH THE NASAL PASSAGE

Dolphins have adapted very well to sending and receiving sounds underwater. Their sonar signal transmission system is like a flashlight with a point light source, a reflector, and a lens. A dolphin does not have vocal chords; it generates sound in a tissue complex that includes a pair of fatty pouches embedded in a pair of soft tissue lips located in its upper nasal passage. These fatty organs vibrate as high-pressure air passes between them to produce high-intensity short-duration pulses. Each sound lasts a ten-thousandth of a second or less, and sounds like a click. Dolphins produce series of tens to several hundreds of clicks at a time. Sound that travels backward is reflected forward by the dolphin's skull. The sound that travels forward is focused in a bulging forehead called the "melon" organ, which is an acoustic lens that dolphins use to focus outgoing sounds.

Underwater sound is received through the lateral side of a dolphin's lower jaw similar to the way in which we hear underwater sounds through bone conduction. While the dolphin's earhole does not function as a sound-receiving pathway, studies have confirmed the high sensitivity of the jaw area. Look closely behind a dolphin's eye to find the small earhole, it is completely plugged by earwax. Sound travels through the acoustic fat inside the lower jaw to the bone that surrounds the dolphin's inner ear system. The inner ear system is similar in all mammals, including those that live in the sea. Toothed whales, such as dolphins and porpoises, have inner ears designed to detect high-frequency sounds, while those of baleen whales are more suited to low frequencies for communication over much longer distances.

How Echo Sounders Work

When you are out on a mountain hike and shout across a valley, you can hear the echo of your words return from the opposite side after a few seconds. If you clap your hands hundreds of feet away from a tall office building on a quiet early morning, you may hear the echo come back from the wall.

Since sounds that are audible to the human ear are reflected by large obstacles, you will not hear an echo from your coffee cup, even if you shout into it. To be able to listen to the echoes from a coffee cup, you need to use a higher frequency of sound.

COPYING DOLPHINS

When it comes to finding fish, dolphins certainly know how to do it best, and we use the same frequencies that they do. We have selected frequencies of 38,000 to 200,000 oscillations a second; one oscillation per second is called 1 hertz (Hz). These frequencies are much higher than the human voice, which is below 1,000 Hz at the dominant frequency. Small targets such as fish require higher frequencies to bounce the sound back. This is also why hospital ultrasound units use a much higher frequency—millions of hertz—to see whether an unborn baby is going to be a girl or a boy. On the other hand, very large targets, such as a building or a mountain, reflect much lower frequencies. So to detect a large submarine underwater, only 100 to 500 Hz in an antisubmarine sonar system would be required. Using the right frequencies to obtain echoes from targets is vital, which is why our fisheries' echo sounders copy dolphins.

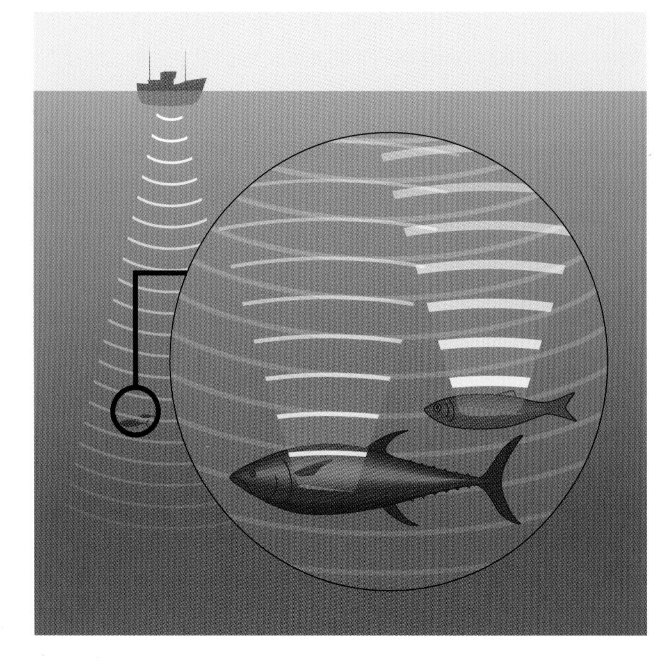

◀ The echo reflected from a fish comes from its swimbladder and has nothing to do with the fish's size. A small fish may therefore reflect a big echo, and vice versa.

DISTANCE OR DETAIL?

The distance that different frequencies can travel is very important. In general, higher frequencies die out more quickly and so do not travel as far as lower frequencies. Therefore a medical doctor has to push the sound probe of a high-frequency hospital ultrasound imaging system close to an organ in order to examine it. This is sufficient to see inside the human body, but useless for detecting submarines, even if it does provide very fine-resolution images. The low-frequency sonar used by navies for antisubmarine detection uses a very loud sound that has to travel extremely long distances and mutes the echoes of small objects such as fish. Higher frequencies give better resolution, while lower frequencies give greater range. Dolphins' and fisheries' echo sounders use frequencies of 100 kilohertz (kHz), a good compromise between the size of the fish that can be detected and the practical detection range, namely several hundred feet.

STEALTH FISH

Bigger fish do not always produce bigger echoes. The bluefin tuna, for example, is a big fish that is relatively invisible acoustically. Echoes from fish reflect mostly off their swimbladders because the water-flesh boundary acts as a good reflector of the sound. A swimbladder is a large sac with thin walls. It acts as a float, and as a sound receiver in some fish species. The juvenile bluefin tuna has a very small swimbladder that renders it difficult to detect by echo sounders, and the same applies to squid or flatfish because they have no gas-filled organs at all and are poor sound reflectors as a consequence.

▼ This shoal of bluefin tuna would be very hard to "see" acoustically. The swimbladder in these fish is very small and therefore difficult for echo sounders to locate.

Filtering and Interpreting Sounds

Noise reduction is extremely important in the practical use of echo sounders, and much work has been done to increase the sound level, augment the receiver sensitivity, and improve the noise filter. While conventional echo sounders are reliable in many situations, they also have limitations.

When fishermen are searching for fish with an echo sounder, a major source of unwanted noise comes from the fishing vessel itself. One effective countermeasure is to use tone bursts. A tone burst has a single-frequency component that is easy to extract from the surrounding noise. A band-pass filter is a powerful tool for eliminating background noise. It allows a certain range of frequencies through, while reducing the strength of all the others. This is just like a radio or television that selects only a certain channel. As long as a fisheries' echo sounder uses only a signal with a single frequency, echoes that bounce back should be at the same frequency. Fishermen can focus on the transmitted frequency to measure the echo level from groups of fish and ignore any other frequency components. This is the great advantage of conventional echo sounders for detecting targets that are a long distance away.

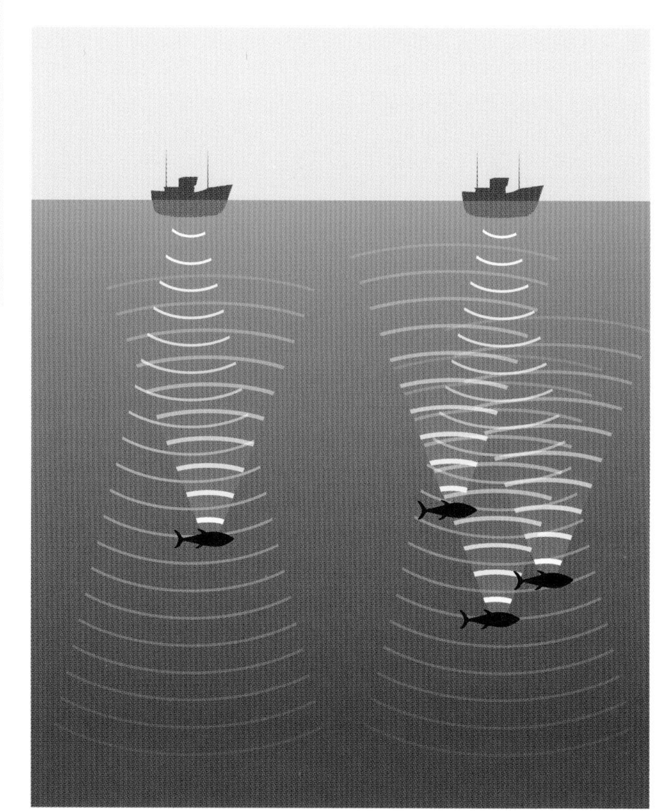

▲ One fish reflects one echo; three fish reflect three echoes. Note that some of the echoes overlap, which makes it difficult to receive them one by one. The conventional echo sounder measures accumulated echo levels and provides the number of fish by dividing the accumulated total by an echo level of a single fish.

COUNTING ECHOES

The number of fish within the acoustic beam can be estimated by the echo level. One fish creates one echo.

◀ Conventional fisheries' echo sounders are very reliable for locating and assessing fish stocks, but only if the fish species and size are already known.

Ten fish create ten echoes. Adding up the energy of all echoes indicates the total number of fish in the projected ultrasonic beamwidth. Once the echo level has been calibrated for a single fish at the transmitted frequency, the number of fish can be calculated. This method is used for the management of fish stocks around the world. The reflectivity of a fish is called the "target strength," which is the ratio of the echo level to the incident sound level. Target strength can be measured using ultrasonic tone bursts of a specific frequency in a test tank. Precise estimation of the target strength is important in assessing the fisheries resources in any specific area of the ocean. This is critical to governments that are trying to manage fish stocks in a sustainable manner. A 10 percent error in target strength estimation causes a 10 percent error in the estimation of the fish resources and has a resulting impact on the determination of the permissible catch. For 1 million tons of sardines, for example, this 10 percent error represents 100,000 tons that can or cannot be captured in a year, depending on the outcome of the measurement of the target strength. Once a target strength is available for each stock and each species, conventional fisheries' echo sounders are very useful for assessing fish stocks, even in noisy environments, as long as the species and size of fish to be caught are known.

HITTING LIMITATIONS

Conventional echo sounders have disadvantages as well. Pure tone bursts are not particularly useful for classifying targets because the echoes contain limited information. We can measure the intensity of the echo only at the operating frequency. A loud or faint echo simply indicates a large or small group of fish. Moreover, the target strength—the acoustic unit related to a single fish—depends not only upon the species, but also on the size, position, and internal structure of the fish. That cannot be estimated without looking at the target fish in the water. This results in the paradox that fish resource assessment is possible only once the fish species and size are known. We want to know the species in advance, however, without catching it.

How Do Dolphins Do It?

The biosonar used by dolphins is a type of broadband sonar. Multiple frequencies are useful for target classification, and these are usually produced by an impulse that contains many frequencies at once.

▲ **Tapping sound is key to discriminating targets in echolocation. It produces vital information, just like tapping a musical triangle will give us a clear, metallic ring, whereas a drum will produce a more hollow sound.**

Tapping on a watermelon is a clear example of using sound to discriminate timbre. Using the same finger to tap different watermelons may produce different timbres. Dolphins and porpoises use ultrasonic short impulses with a broadband spectrum to "tap" remote targets. The echo from a broadband signal contains multiple frequency components representing the target's characteristics, such as the clear tone of a triangle and the hollow note of a drum. Some frequency components survive the reverberation and reflection from the target, while others do not. This does not happen when the sound is a pure tone burst because the echo will contain no other frequency. A broadband signal is thus more useful for determining the characteristics of a target.

An advantage exists to using identical impulse waves. Even with a series of identical impulses, the echoes have different tone characteristics depending on the target. Try tapping on a laptop computer with a pencil, then on a computer mouse. The tone of the echo is different for each target. If we tap the laptop with a pencil and then with a hammer, the resulting sound will be completely different, even for the same target, although the computer may be damaged by the second choice! Transmitting a series of identical impulses and listening to the response from the target provide a convenient method of classification. Dolphins and porpoises produce sequences of similar clicks, which bounce off their targets, especially prey, and return as echoes that are translated into information about the size, shape, and distance of the target. This is just like tapping a target many times using the same pencil.

BIOSONAR EXPERTS

Let us see how dolphins are biosonar experts. Using very high-intensity and short-duration impulses, a bottlenose dolphin can discriminate target thickness, material composition, and even the shape of a target. Trained

▲ Dolphins are able to "see" in the dark depths of the ocean simply by emitting clicks and bouncing them off different targets.

dolphins can provide a simple go or no-go response by touching a paddle when presented with a target it has been trained to recognize. Extensive studies performed by the U.S. Navy have demonstrated the amazing abilities of dolphin sonar. The bottlenose dolphin can detect the existence of a metal target sphere 3 inches (7.62 cm) in diameter at distances of up to 370 feet (113 m). Moreover, it can also detect a 0.01 inch (0.3 mm) difference in the thickness of the cylindrical target.

If you were to touch two identically shaped wine glasses having slightly different thicknesses, would you be able to tell them apart? Sound would work better than touch in this case. Tap each wine glass gently with a spoon and listen … you will hear a clear difference in tone. Dolphins can detect subtle differences in materials, easily differentiating between bronze, aluminum, or coral, for example. Interestingly enough, once the dolphin's echo frequency is shifted to the human audible range, humans, like dolphins, can also discriminate between different material compositions by listening to the echo.

CHECKING DISTANCES

Another singular feature of the dolphin impulse sonar is its high spatial resolution. The duration of a dolphin's click is very short. During its 50- to 100-microsecond duration, the sound travels 3 to 6 inches (7.5 to 15 cm). This means that the echoes from two targets separated by more than 6 inches (15 cm) can be discriminated. This ability is quite useful to dolphins as they feed. They need to focus on individual fish because they catch their prey one by one. This method of feeding is unlike that of baleen whales, which are not believed to have the ultrasonic sonar ability possessed by dolphins. A baleen whale simply opens its mouth and takes in small fish, squid, and zooplankton along with seawater; when the whale closes its mouth, the baleen acts like a strainer to hold the prey while releasing the water.

Manufactured vs. Dolphin Sonar

Dolphins have binaural hearing: in other words, they have two ears to locate sound sources, just as humans do. Both dolphins and humans can identify the direction of sound sources very precisely. They, and us, are able to use our two inner ear systems to locate and tune into softer sounds in noisy environments.

You can test this ability quite easily. Ask a friend to stand directly in front of you and clap. Then close your eyes, and ask your friend to move only one step to either the left or right and clap again. You can easily tell to which side the person moved. The difference in the time of arrival and the received intensity of the sound at your two ears are cues to identifying the direction. The location accuracy of both humans and dolphins is within 1–2 degrees.

FOUR RECEIVERS

A manufactured split-beam echo sounder usually has four channels to locate the sound source direction in two dimensions by using differences in sound arrival times and intensities between channels. One pair of receivers provides one bearing angle. Two pairs of receivers indicate the specific direction of the echo. The remaining dimension, the depth of the fish, is the easiest parameter to measure, by calculating the delay time of the echo from the sound transmission. A split-beam echo sounder can hence measure the three-dimensional position of the target fish.

Evolution has not provided dolphins with additional ears; however, a bottlenose dolphin can determine the vertical elevation angle by echolocation. This may seem theoretically impossible at first because sound arrival time and intensity should be identical at the left and right ears as long as the sound source is on the center line of the dolphin.

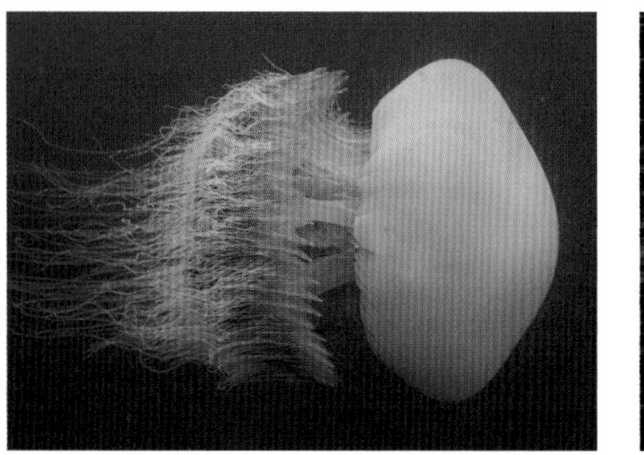

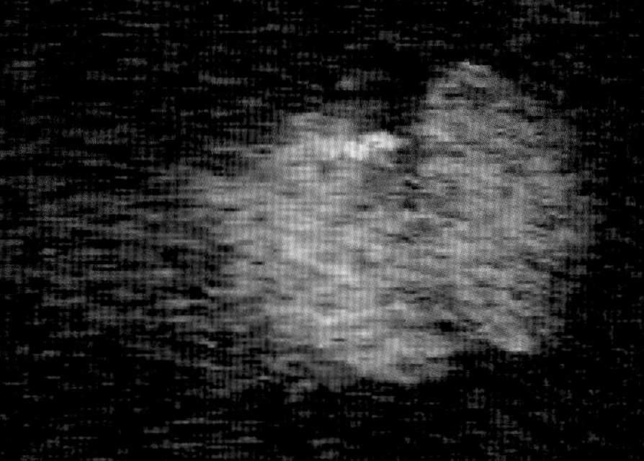

◀ Visual image of a Nomura's jellyfish (*Nemopilema nomurai*, left) and (right) acoustic image captured by the DIDSON acoustic camera. Images were taken in Wakasa Bay, Sea of Japan (courtesy of Naoto Honda, Fishing Gear and Method Laboratory, National Research Institute of Fisheries Engineering, Fisheries Research Agency).

ULTRASOUND IMAGING

A pregnant mother can see her unborn baby through an ultrasound imaging system. This system uses very high-frequency ultrasonic waves at millions of oscillations per second to obtain a detailed image of the fetus. The array of receivers in the ultrasound probe is just like an acoustic lens. Multiple arrays of receivers enable identification of echoes from any direction to extend the horizontal dimension of the image. This technology enables us to see inside the womb, ascertain the sex of the fetus, and check for abnormalities.

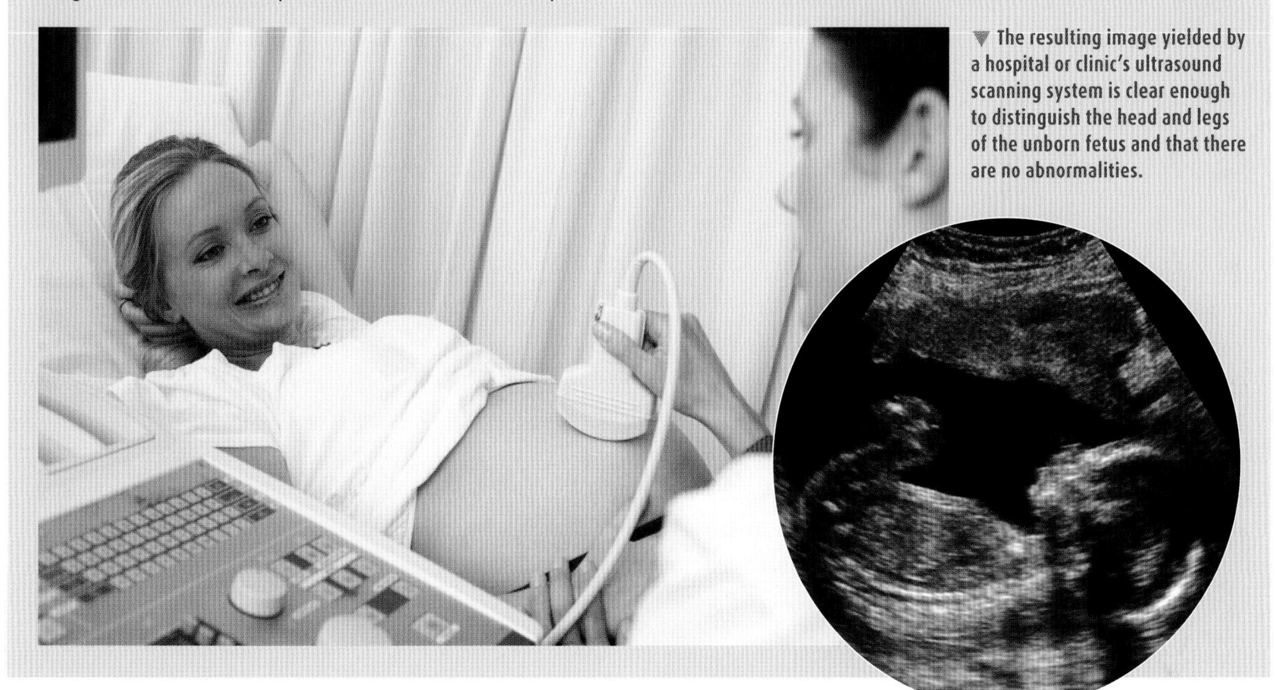

▼ The resulting image yielded by a hospital or clinic's ultrasound scanning system is clear enough to distinguish the head and legs of the unborn fetus and that there are no abnormalities.

Additional acoustic cues are needed to identify the elevation angle. The dolphin's head is symmetrical left and right, but not up and down. The sound characteristics are modified as the sound travels through the head of a dolphin. The sound the dolphin receives from above is slightly different to what it receives from below. As long as the same impulse sounds continue, the echo characteristics could indicate the vertical angle.

Multiple echoes from the underside of the water surface are another possible cue. The mixture of direct and reflected echoes depends on the position of the target relative to the dolphin, in the same way that multiple echoes in a concert hall provide a special "acoustic scene." If an orchestra plays the same music outside, it sounds completely different.

ACOUSTIC IMAGING

A digital camera has a lens to focus the image on the imaging element. In a similar way, the DIDSON underwater acoustic camera developed by Sound Metrics Co. has an acoustic lens to focus and form an acoustic image. The DIDSON provides a visual image of objects even in murky or completely dark environments.

This underwater camera works in rocky, uneven ocean-bed environments where other equipment has failed. It can distinguish between different underwater plants and sea creatures with more accuracy, and works well in underwater high-security surveillance operations. It is also an effective tool in pipeline inspections and maintenance, and plays an important part in leak and flow identification.

Perceiving Internal Structures

Dolphin sonar sounds propagate only in water—they do not travel in air—so dolphins are not able to examine objects acoustically when they are out of the water. In spite of this, one dolphin was successfully trained to match a visual image of a target in the air with the acoustic image in the water.

The object was then submerged in an optically opaque plastic box that was transparent to echoes. After an acoustical examination of the box, the trained dolphin correctly positioned itself in front of one of the correct objects presented visually in the air. This suggests that the dolphin can extract object shape directly from echoes. But how could the dolphin recognize the structure of the target?

EYES AND EARS

The human eye has two types of detectors called "cones" and "rods." Cones sense colors, and rods are sensitive photoreceptors for low-intensity light levels. These detectors are distributed over the retina. The image through the eye's lens is projected onto the retina so that two-dimensional information of the target can be obtained. This is similar to the digital camera or the DIDSON acoustic imaging system. Dolphins and bats, however, have only two ears. While you may be able to tell the horizontal direction of a piano, cymbals, and violin in an orchestra just by listening, you cannot "hear" the shape of the piano, cymbals, or violin. Our binaural hearing system is not sufficient for obtaining two-dimensional images of a target just by listening to the sound from the object.

Researchers have been trying to explain this mystery. An object of a certain thickness reflects sounds from the

◀ Dolphins cannot use their wonderful sonar abilities in the air because their sonar sounds do not travel well when out of the water.

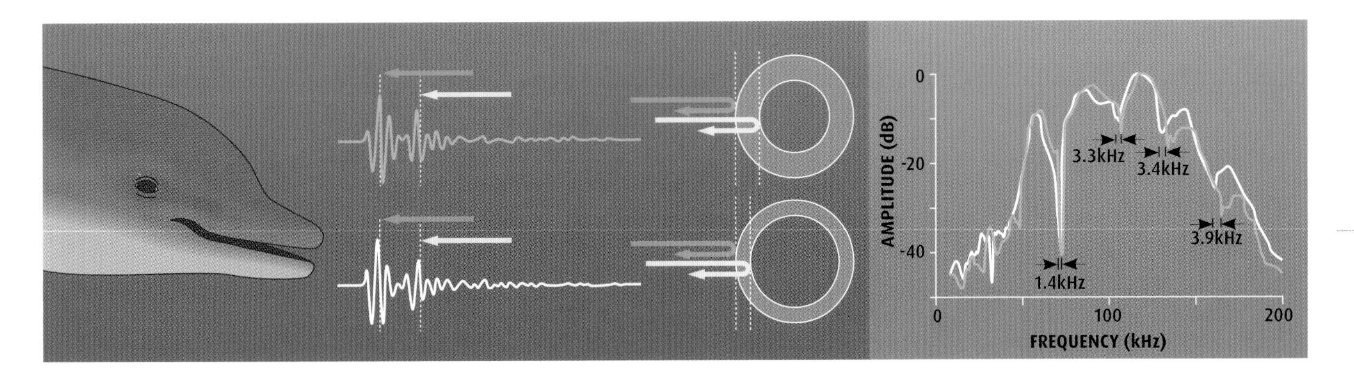

▲ Echoes from a standard cylinder and a comparison cylinder, which was 0.01 inch (0.3 mm) thinner than the standard one. The double pulse structure of the echoes from the outer and inner surfaces of the cylinder caused several notches in the power spectrum shown in the right image. The difference of the thickness corresponds to the difference of the frequency at the notch.

outer and inner surfaces, so that each single impulse results in two echoes separated by a time related to the object's thickness. The frequency spectrum of two successive pulses mathematically shows an up-and-down shape of spectrum components according to the frequency. The separation of the notch in the spectrum corresponds to the thickness of the target.

LOOKING INSIDE

The time lapse of echoes has been proposed as one way of determining the internal structure of a target. One researcher measured the echo from a fish using a sonar signal mimicking that of a dolphin. The echo produced had several impulse components that indicated multiple targets inside the fish's body, including the swimbladder, the head, and possibly the spine, and dorsal skin. The arrival time of each echo was measured and used to calculate the geometry of the fish's internal organs, which closely matched an X-ray image of the same fish. This information was sufficient to tell whether the fish was approaching or departing because the echo structures from the front and back of the fish were different.

This system could be used in a future fish direction detector similar to the ultrasonic imaging system that shows the direction of blood flow in a human heart. It is clear that we have a great deal to learn about the amazing sonar capabilities of dolphins. Moreover, once depth information is available from a specific direction, a dolphin can swim around the target and scan it from different directions and hence three-dimensional reconstructions may be possible. This is, however, currently mere speculation. We still have a long way to go in order to understand the mysteries of dolphin sonar recognition.

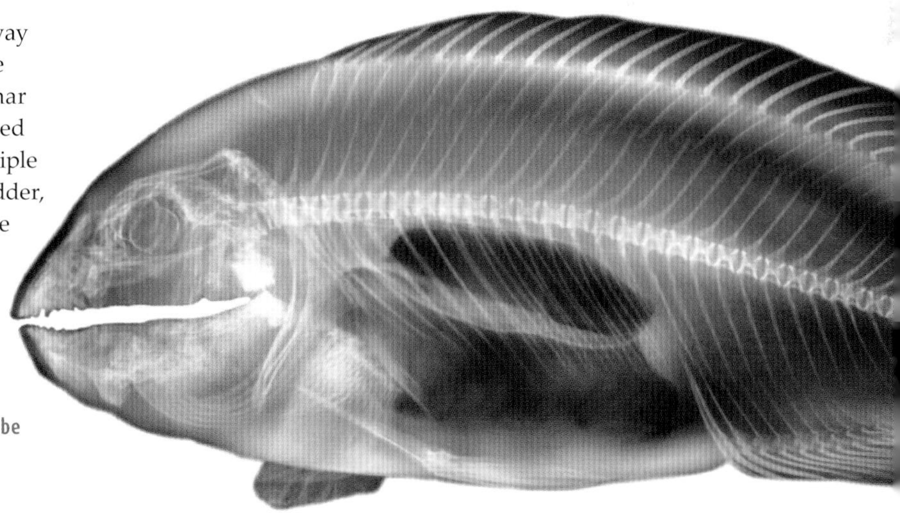

▶ X-ray image of a fish. A swimbladder is the major sound reflector; however, the head, skull, spine, and even the skin could be minor reflectors of a sound. The echo structure, depending on the position of each reflector, could be a key to classifying targets.

Sharpening Sonar Focus

Sharp focus is vital for obtaining clear target images, and optical focusing is a well-established technology for still and video cameras. Early cameras required manual intervention to focus, but modern compact digital cameras have autofocus and even automatic human-face recognition.

Once the lens focuses on a face, for example, the camera will capture a sharply delineated image, while surrounding objects may be blurry. Light from the contours of the face is refracted through the lens and forms the same contours on the camera's screen. Light from objects at different distances is refracted on different paths by the lens, and even points of light are scattered on the screen if it is out of focus.

TIME DELAY

The sonar of dolphins and porpoises has a focusing function, but the mechanism is different. These creatures adjust the timing of the sound transmission instead of controlling the refraction pathways to focus on an echo. Dolphin sonar signals have many varying ultrasonic pulse sounds. A dolphin transmits the second pulse only after receiving the echo of the first. Note that the speed of sound in water is approximately 1,640 yards (1,500 m) per second. If the target is 16 yards (15 m) away, the dolphin waits at least 0.02 seconds to produce the next sound because the sound takes that long to make the 32-yard (30-m) trip to the target and back. Dolphins actually wait a little longer, and this extra time is used for processing the echo signal. Dolphins are able to determine the distance to the target by measuring the time required for the echo to return after each pulse. This is reminiscent of a ball game. A boy throws a ball to his friend. His friend in turn throws the ball back to him. The boy is then able to throw the ball to him again. Each

◀ **Just as children in a ball game will wait long enough for a ball to travel back and forth for them to catch and return it, a dolphin waits just long enough for an echo to travel back and to be processed, to avoid "jamming" it with the next one.**

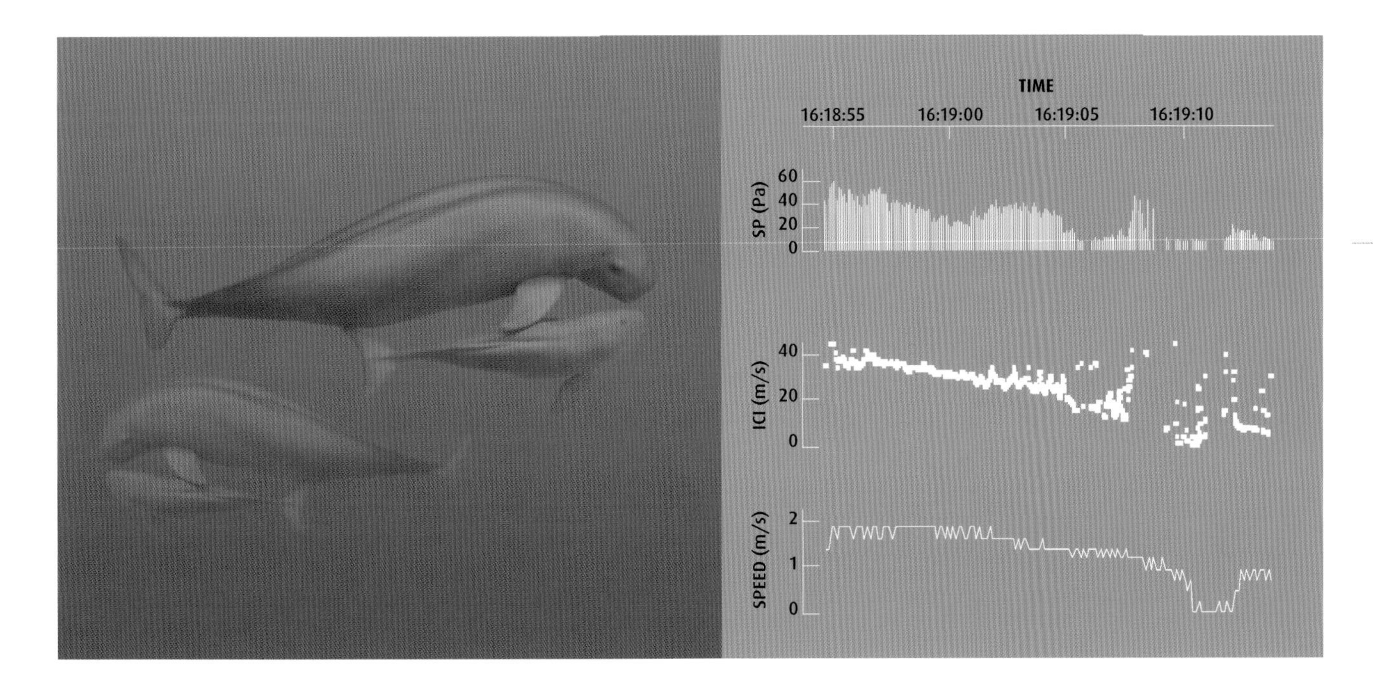

▲ A finless porpoise reduced the inter-click interval (ICI) and sound pressure (SP) as it approached a potential target. In the meantime, the time delay of the echo from the target was reducing. The porpoise kept moving, which was measured by a speedometer attached to the animal.

boy does not wait for a long time with the ball in his hand; they each need only a couple of seconds to be in a set position after catching the ball. If the boys have two balls, the timing becomes complicated and they are unable to play the game. Similarly, a dolphin waits for the echo to return to avoid jamming it with the next one—like hitting two balls together. Dolphins "throw" tens, or sometimes hundreds, of ultrasonic pulses to hit a target, but process them one by one.

If a boy keeps playing catch with a ball, but walks toward his friend, the time to catch the next ball grows shorter. As a dolphin approaches a target, it reduces the interval to the next pulse production according to the target range. In other words, the dolphin controls sonar focus range according to the distance to the approaching target, just like the autofocus camera.

DOPPLER SHIFTS

Bats use Doppler shifts to measure the relative speed of their prey. A Doppler shift is heard in the change in tone frequency of an ambulance siren as it approaches, passes, and then moves on. The siren sounds higher when it is approaching and lower after passing, and the shift in frequency depends on the speed of the approaching/disappearing sound source. Bats produce tone bursts similar to those of a conventional echo sounder. The echo reflected from an approaching insect has a higher frequency than the one originally produced by the bat. If the insect is flying away, the tone will be lower. Knowing the relative speed of, and distance to, the prey helps the bat to predict the position of the target a few seconds before capturing it.

Dolphins and porpoises do not seem to use Doppler shifts for speed measurements. Their pulses are too short to detect the tiny shifts. Instead the creatures find the trajectory to a target by repeatedly bouncing echoes off it.

Toward Dolphin-Mimetic Sonar

Major differences between dolphin sonar and conventional fisheries' echo sounders are the broadband frequency characteristics and high spatial resolution of the former. Engineers and scientists are currently trying to develop new-generation echo sounders with such capabilities.

A major fisheries acoustics company called SIMRAD, based in Oslo, Norway, has developed a multibeam echo sounder called the ME70. This system operates in the 70- to 120-kHz frequency range, not at a single frequency. The frequency response of different sizes and species of fish can be identified using the broadband frequency response from the fish. SciFish model 2100, from Scientific Fishery Systems, Anchorage, Alaska, uses a 60- to 120-kHz frequency to achieve a broadband response from fish. The device was tested on free-swimming alewives, rainbow smelt, and bloaters in the Great Lakes of North America. A computer neural network was set up to learn the echoes

DOLPHIN SONAR SIMULATOR

The broadband echo sounder, which simulates dolphin sonar, can distinguish between fish with much greater accuracy than conventional echo sounders can. Simply by zooming in on the images of a school of Japanese anchovy (*Engraulis japonica*) taken by a conventional echo sounder and the dolphin sonar simulator (right), we can see that each trajectory of swimming fish can be recognized by the dolphin sonar simulator, but the conventional echo sounder gives only a limited-resolution image and cannot discriminate between each fish.

COMPARISON OF ECHOGRAM IMAGES OF JAPANESE ANCHOVY

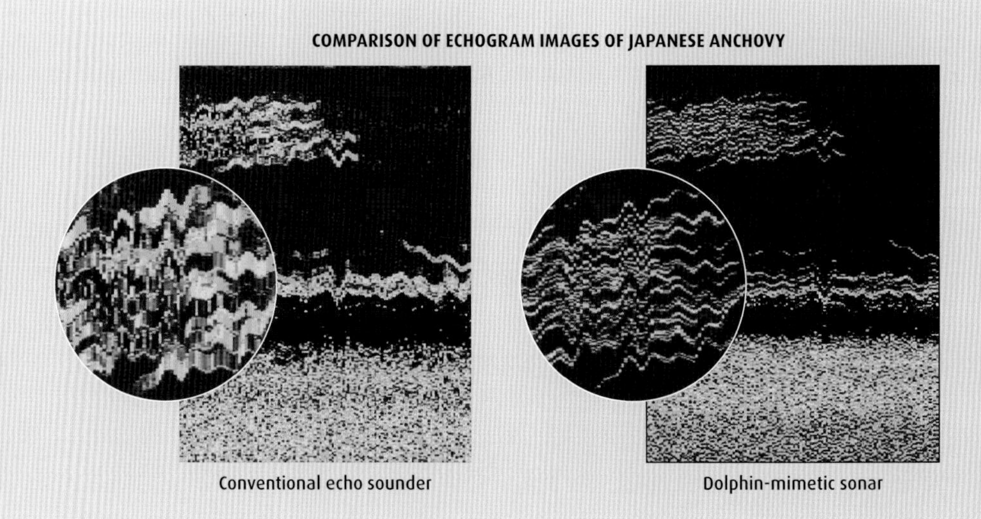

Conventional echo sounder

Dolphin-mimetic sonar

▶ We are currently learning to distinguish between sea creatures in the same way that dolphins do. For example, we now know that squid reflect echoes comprising a single highlight, with an occasional softer, secondary highlight.

produced by different species. Tests show that the device could identify some lake fish with 80 percent accuracy.

Broadband dolphin echolocation signals were used to measure the backscatter (echoes that travel away from the source of the sound) of marine creatures. The echoes from myctophids (lanternfish) and shrimp usually had two highlights: one from the surface of the creature nearest the echo sounder and a second that was probably from the signal propagating through the body of the creature and reflecting off its opposite surface. The squid echoes consisted mainly of a single highlight, but sometimes had a low-amplitude secondary highlight.

The very high resolution of a short-pulse echo sounder worked well in distinguishing between individual fish in a school of Japanese anchovy. This system, called a "dolphin sonar simulator," projects sonar signals of a bottlenose dolphin and has a broadband transducer and receiver adapted for 70 to 120 kHz. Even in a dense group of the fish, many distinctive traces could be recognized by the dolphin-mimetic sonar, whereas the conventional echo sounder provided a relatively low-resolution image of the fish school. Using high spatial resolution and broadband echo characteristics, fish size and species identification will be possible even in a dense group of mixed fish.

CONCLUSION

Accurate classification and identification of underwater targets is vital, not only for fisheries surveys, but also for underwater security measures. To achieve this, researchers have recently attempted to mimic dolphin sonar. The next step is to identify key acoustic characteristics, which are probably embedded within each echo, to distinguish one target from another. Analyzing spatial and frequency structures of echoes from different types of targets will be helpful.

Broadband sonar is more strongly affected by noise contamination than conventional narrowband sonar is, and countermeasures for noise reduction are still poor, so engineers will need to improve broadband transmission efficiency and reduce noise produced by ships. To make broadband sonar viable as a commercial product, we need to overcome the current system's large size and huge energy consumption. In the meantime, development of a broadband sonar system that is able to find and classify invisible targets will require more research on the mechanisms of dolphin sonar, which have evolved over time to ensure dolphins' survival underwater. Fortunately, elementary technologies for simulating dolphin broadband sonar are now available due to the rapid progress of technology. Broadband fine-resolution sonar, inspired by dolphins, is therefore set to become an invaluable tool for "seeing" underwater objects in the future.

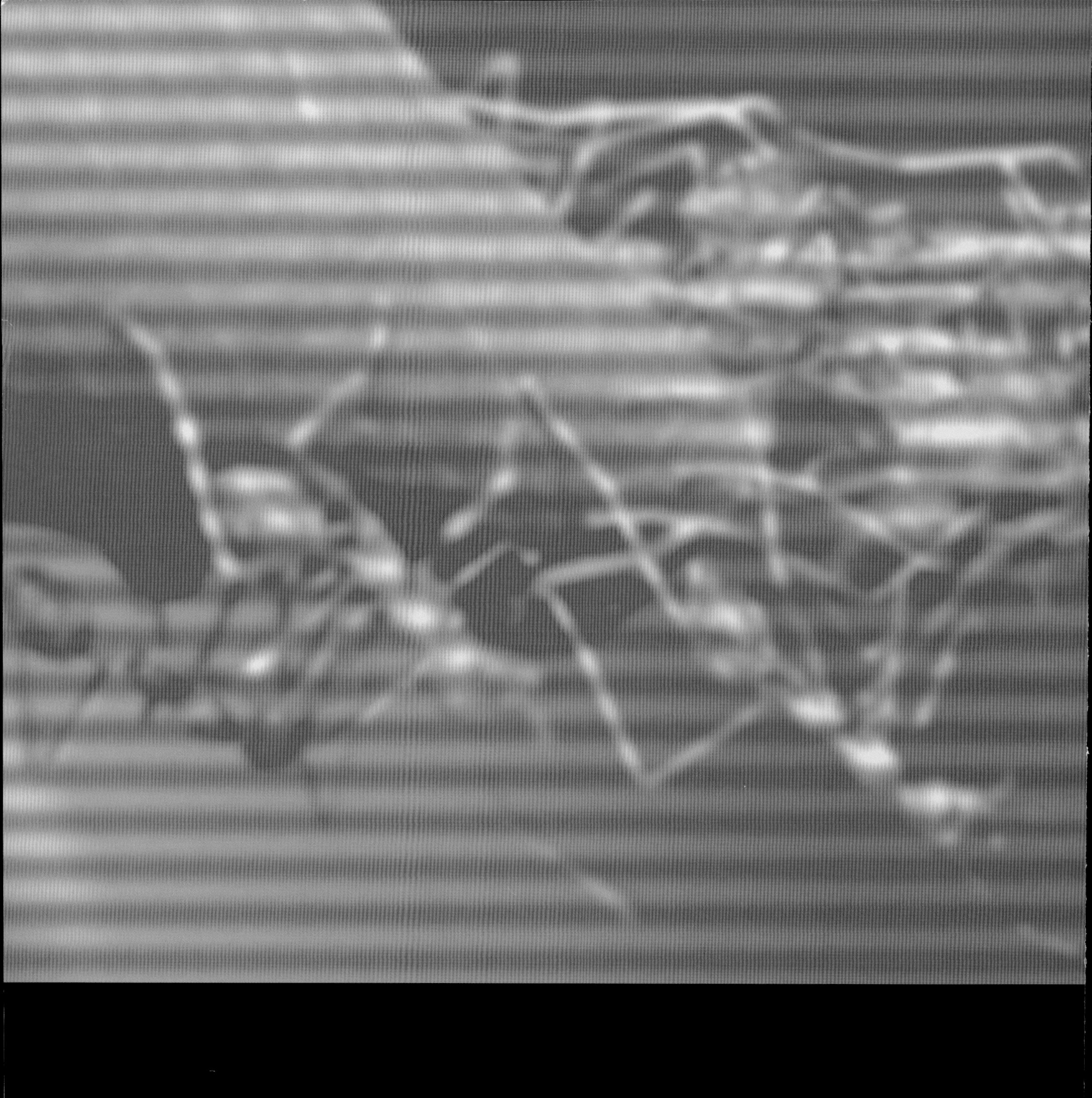

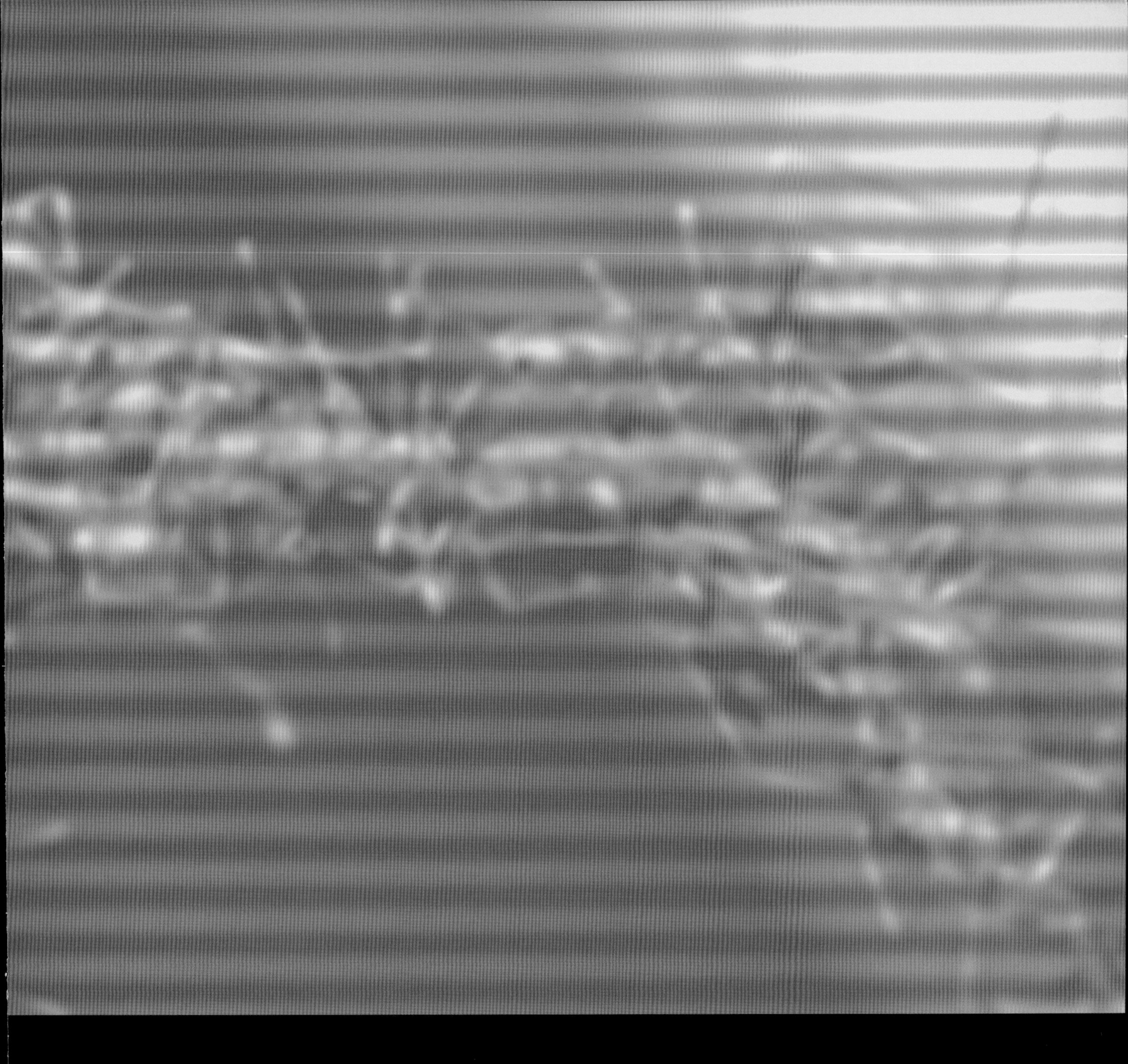

4 | COOPERATIVE BEHAVIOR

Introduction

Cooperative behavior is seen in many successful groups of animals, and such behavior reaches a pinnacle in the colonies of social insects such as bees and ants. Their distributed sensing, synthesizing, and control systems have produced some of the most socially advanced non-human organisms on the planet.

So successful are they that, despite representing a very small proportion of known insect species (approximately 2 percent), they compose more than half the biomass. Ants alone have roughly the same biomass as humans—definitely a thought-provoking statistic. Superb and inspiring these insect colonies certainly are, and such successful groups can offer insights into self-organization and cooperative behavior that we could use in the design of engineering systems such as teams of robotic vehicles and beyond to improve on our current methodologies and techniques, and to gain considerable efficiencies of operation. This chapter offers an account of how this might be approached and some of the lessons learned to date. To begin, let us look at some of aspects of cooperative behavior and the range of animals in which this has evolved to a high level of sophistication.

◀ **Teamwork enables climbers to achieve greater heights than could be attained in isolation.**

▶ Landing people on the Moon or orbiting the Earth in the space shuttle is possible only through sophisticated team organization.

PART OF A TEAM

Understanding human group behavior—for example, interactions and reactions to economic events and their consequences for financial markets—is clearly a valuable and laudable objective to aid prediction of potentially large fluctuations. Cooperative behavior is fascinating in its own right and is becoming increasingly studied, but a major interest for our purposes is the organization and management of teams. As we move in the field of robotics, say, from the control of individual devices into the development of team working, we may be able to learn from the cooperative behavior of animals and, hopefully, find a shortcut to the implementation of successful strategies. This is particularly relevant to robotic vehicles, where the trend now is to replace single, expensive vehicles with a team of small, low-cost vehicles. A team has the potential to cover an area of exploration more quickly and also to create built-in redundancy and robustness to failure. If a vehicle is lost, the team can still achieve its mission. The problem we now have is how to organize the team to achieve the best performance.

The world is full of success stories of team working when the brainpower of the team members is harnessed, combined, and channeled into a project. Landing people on the Moon became possible, even with the limited computing power available in the 1960s, and unbelievable engineering construction projects such as the Channel Tunnel linking England and France have been undertaken and completed despite seemingly insurmountable obstacles. Standing on the summit of Mount Everest in 1953, Hilary and Tenzing were absolutely dependent on the pyramid structure of a large team that established camps and transported food and equipment to strategic points along the route up the mountain. The management of such a team is crucial to success and was, no doubt, one of the main reasons why this team succeeded where others had failed. A clear structure of delegation of responsibilities is essential.

Humans have considerable brainpower available, which we are beginning to understand in more detail, but the network complexity involved actually works against using similar networks for engineering systems. Nevertheless, we are beginning to use very simplified versions of neural networks in order to capture some of the advantages of information synthesis and simple decision-making. Perhaps, however, we could learn more about developing teams of cooperative engineering systems by looking at animal group behavior that has clearly led to very successful animal species.

Safety in Numbers

The idea that fish congregate in shoals is so familiar from nature programs on television that we tend to think of it as completely unremarkable. But it may be that shoals of fish are so commonplace that we never ask the question: Why do fish shoal? There are several possible reasons.

▼ **A shoal of fish provides safety in numbers and social interaction, while schooling helps swimming efficiency because the fish in the inner part of the shoal encounter less water resistance.**

Safety in numbers means that, when a predator attacks, the chance of an individual being captured is lessened. Multiple sense organs considerably enhance awareness and the readiness level to meet a possible attack. Social interaction is also a key driver in shoaling, as is the increased opportunity for reproduction. Detection of food sources is enhanced, but so is the level of need. There is the additional problem that fish on the outside of a shoal may be able to eat, but those on the inside may find it more difficult. Life on the inside may be safer, but those individuals must move to the outer layers to feed and hence increase the risk of predation. Life still presents many problems, even for a shoal, but nevertheless, the advantages clearly outweigh the disadvantages, which is why about a quarter of fishes shoal all their lives, and about a half of fishes shoal for part of their lives.

Schooling is generally taken to mean swimming in the same direction in a coordinated manner, such as is seen in a dramatic way in the incredible numbers of migrating sardines. This can lead to improved hydrodynamic efficiency in the same way that geese fly in a V formation to gain aerodynamic efficiencies from wingtip vortices. Group behavior can lead to stunning patterns of movement, as seen in fish schools when predators attack. Similar patterns can be seen in flocking starlings, which present amazing aerobatic displays as they return to their roost sites.

Can such group behavior hold lessons for the design of technological systems? Let us go underwater to look at just one example of where this is already showing promise.

◀ Flocking starlings perform stunning aerial display patterns before returning to their roost sites. Scientists are studying this.

▶ Autonomous underwater vehicles can explore areas too dangerous for manned submersibles, but are very expensive to produce.

EXPLORING THE OCEAN ENVIRONMENT

The seas cover approximately 70 percent of our planet, and there is much to explore. It is said that we know more about space than we do about the ocean. Exploration of the underwater environment is exciting and important, not only to develop our understanding of the environment and the life it supports, but also to be able to harness sustainable sources of energy. The economics of underwater exploration and the need to reduce potential risks in very deep waters have led to the development of unmanned underwater vehicles (UUVs). Autosub is one example, a pioneering vehicle developed in the UK over many years by the Natural Environment Research Council. Autosub has been used in exploratory missions ranging from looking at landslips that might lead to earthquakes in deep-sea areas to studying the effects of climate change under the Antarctic icecap. Working under the ice poses considerable safety problems for manned submersibles and is very hazardous even for unmanned vehicles. The technical challenges are extremely demanding for vehicle designers, and losing a vehicle would be very expensive.

SUPERSUB

Autosub is 23 feet (7 m) long and 3 feet (almost 1 m) in diameter, with a range of more than 300 miles or 6 days, depending on the energy requirements. To date it has undertaken almost 300 missions, with the longest covering more than 160 miles (257 km) in 50 hours. It has traveled unescorted for more than 1,200 miles (1,931 km), with the deepest dive being down to more than 3,300 feet (1,000 m). Impressive statistics. Launch and retrieval of Autosub is from a research vessel and is an expensive business, but the physical, biological, and chemical sensor payloads and the operational envelope of the vehicle mean that the data acquired are not only extremely valuable, but also impossible to obtain by conventional methods. Such data are used to study underwater geology, ocean biochemistry, ecosystems, and ocean temperature, which drives climate.

Control of the vehicle's heading and depth is achieved through rudder and stern plane control surfaces that are positioned by the onboard autopilot, which also drives the propellers and variable buoyancy system. Autosub is largely autonomous when underwater because it has no connecting wires to the mother ship. To determine its position, it has to surface in order to communicate with a satellite from which point it can adjust course if necessary and dive again. Since Autosub is a single sensor platform, however, there are risks such as sensor failure, problems with vehicle subsystems, or even loss of the vehicle itself.

More Is Better

Vehicles such as Autosub (see page 93) are important to underwater exploration, but the cost of developing and running them presents a barrier to their use. Ways are being sought to develop lower-cost vehicles that can be reproduced easily and assembled into teams that can, for example, survey an area more efficiently.

Such a vehicle is Subzero. This is a torpedo-shaped vehicle 3 feet (90 cm) in length and 4 inches (10 cm) in diameter. It has a cylindrical hull, made from Perspex, and removable nose and tail sections. Propulsion is from a 250W 16,000 rpm samarium-cobalt DC motor that is powered by a 9.6V Ni-Cad battery pack. The four control surfaces, linked rudders and two independent stern planes, are actuated by model aircraft servos. The vehicle is directed by a computer on the shore that communicates via an electrical cable using the built-in serial ports of two Motorola 68HC11 microcontrollers, one on board the vehicle, the other operating within the control computer on a custom-built communications card. The communication cable transmits the commands and the sensor data between the shore computer and the vehicle bidirectionally. This system is being replaced by an acoustic communications system, to allow the vehicle to run freely. Basic research has been carried out on control schemes for this vehicle that are highly complex and usually related to just one or two axes of motion. Nevertheless, the development of all the subsystems could be used to create even smaller vehicles that could in future form collaborative UUV teams. The cooperative behavioral aspect of controlling a team of UUVs is a more difficult problem, and this is where biology may hold clues to the best way forward. Before looking at lessons from biology, let us look at some options open to us through the more conventional routes.

▼ Subzero is a small, low-cost vehicle that could be duplicated to produce a team, but we still need to develop a way of coordinating its group behavior.

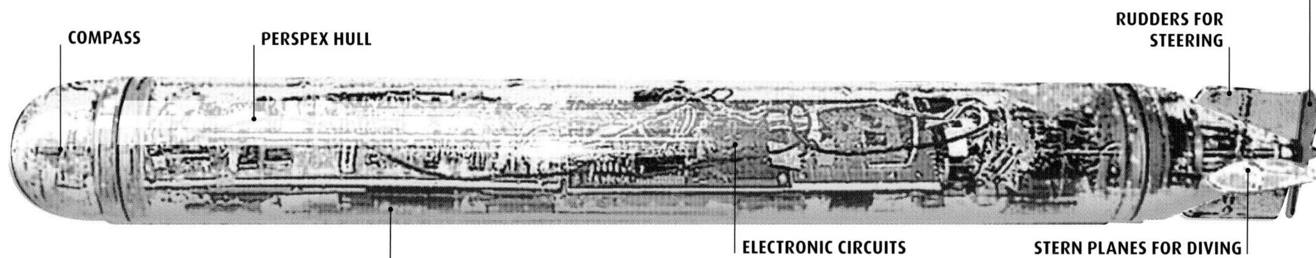

▼ A single vehicle explores an area of seabed in a so-called "lawn-mower" pattern, whereas a team could cover the same area more efficiently.

▶ Controlling a team of aircraft in close formation presents extraordinary challenges. A leader–follower organization works well.

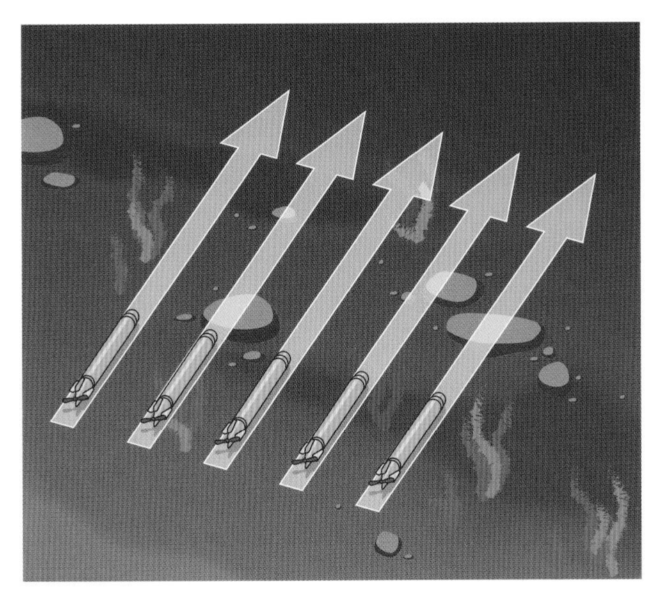

WHO IS IN CHARGE?

A single vehicle would have to, typically, run in a so-called "lawn-mower" pattern in order to comb an area, whereas a team could "fly" in formation and could cover the same area much more quickly. The problem then is how we could control such a team. Control of a single vehicle is a difficult challenge in itself, but multiply this up to the level of a team and we have a major design problem on our hands. Individual vehicles within a team would still have to be able to control their speed and attitude, but how they act as a team and how they are organized is a higher-level problem to which there is no easy answer. A leader–follower structure is appealing because it is used to great effect in controlling teams of humans, but it is vulnerable to the loss of the leader. If the lead vehicle developed a fault, one of the followers would have to assume that role or the mission would have to be aborted. This role reversal would bring with it a different set of problems. If we look at the military scenario, there is a command structure that is well defined from the top down. At the head, the commander is supported by team leaders, navigators, and so on, and we could use this structure in our vehicle team. This again would be vulnerable to loss of a key element, however, particularly near the top of the hierarchy. A change of responsibility across the team would lead to very serious problems of rescheduling tasks, communication requirements, and so on. Losing one low-cost vehicle might be a problem, losing a whole team through loss of a navigator is quite another!

Taking Control

The answer to the problem of group control may lie in biology. It is difficult to study group behavior because of the large number of individual animals involved. Simple models have been created to explain some of this behavior and to act as descriptors that may inform the design of novel approaches to traditional problems.

Simple mathematical models of animal aggregations are generally developed from rules such as:

- avoid collisions with neighbors
- remain close to neighbors
- move in the same direction as neighbors

Such a model is Boids (or "birdlike objects"), which was developed by artificial life and computer graphics expert Craig Reynolds (1987) to simulate coordinated animal motion as seen in bird flocking and fish schooling, using very simple rules. The model describes the interactions of individual agents, with behavior constrained by a set of simple rules based upon the positions and velocities of neighboring flockmates. In other words, individuals react to flockmates only within a small, surrounding neighborhood. Realistic patterns of behavior emerge from these rules, as can be seen on Reynolds' website. In addition to providing a useful illustration of flocking behavior, the model can form the basis of autonomous characters in computer animation and games. It has, for example, been used in film animations such as *Batman Returns* (1992), where modified Boids software was developed to simulate swarming bats and flocking penguins. More complex rules can be added to include obstacle avoidance, navigation to a target, remaining within a given area, and so on.

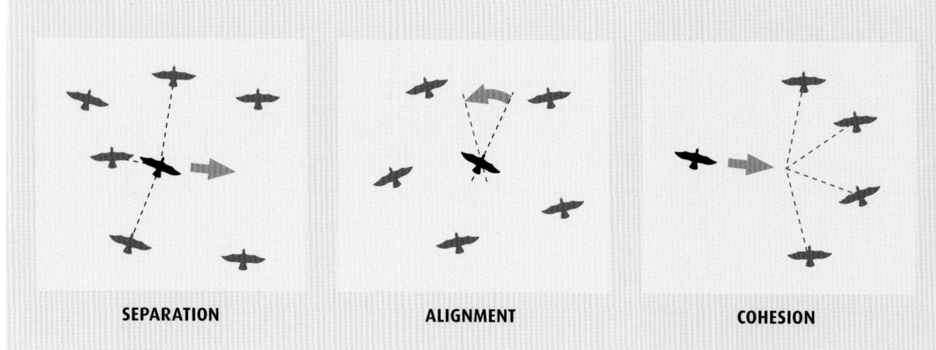

Separation: steer to avoid crowding local flockmates

Alignment: steer toward average heading of local flockmates

Cohesion: steer to move toward average position of local flockmates

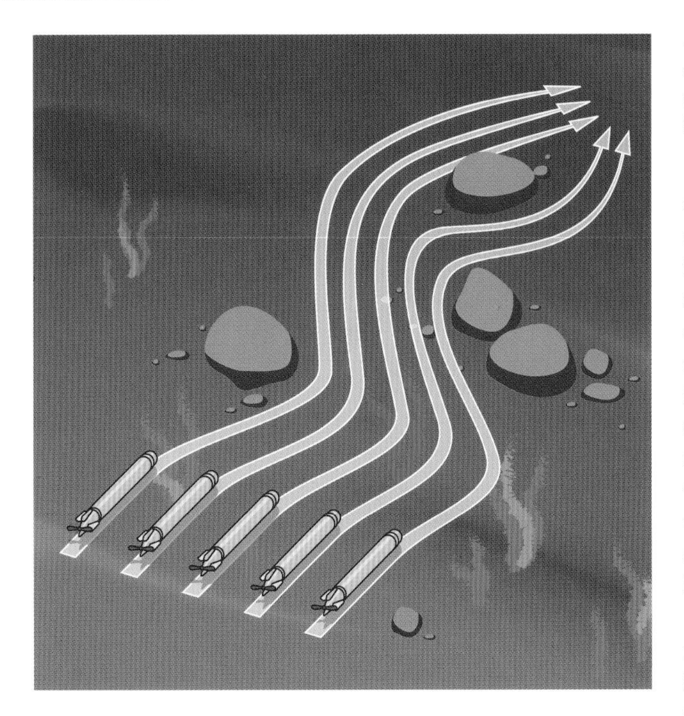

◀ Cooperative animal behavior could help when organizing teams of autonomous underwater vehicles to navigate around obstacles.

ACTING IN CONCERT

If we return to our problem of controlling a team of small underwater vehicles, a computer simulation was created that used an enhanced set of Boidlike rules designed to enable a UUV team to navigate to a target destination, avoid obstacles along the way, and cope with tidal flows. The dynamic characteristics of each vehicle, modeled on Subzero, were included within each "agent" to provide realistic motions following a demanded change in heading.

The picture above shows the typical trajectories followed by each vehicle within a team of five UUVs as they undertake a mission to progress toward a given destination while avoiding obstacles along the way. The vehicles were in formation, as would be the case in some surveying applications in practice. The vehicles begin in a straight line and have to progress to a small rectangular target toward the top right (target not shown). Obviously this is two-dimensional and hence ignores depth changes, but this dimension will be added in the future. The success of the mission is clear, however, and shows considerable promise for the bioinspired approach, which offers a much simpler strategy for navigation and control than more established methods. Naturally sensors onboard the real vehicles would be necessary that work in the same ways as birds use vision or fish use vision and their lateral line, a sense organ that runs lengthwise along their sides and provides hydrodynamic information.

◀ The Boids model uses only three simple rules to explain coordinated motion seen in animal groups such as flocking birds or schooling fish.

In order to take the flocking modeling to a higher level, researchers in the multidisciplinary, multinational StarFlag collaboration are attempting to address the lack of experimental data by trying to determine in three dimensions the fundamental laws of collective behavior and self-organization in animal groups. They have found similarities between bird flocking and fish schooling models generated previously which, when modified to account for bird behavior and interactions, led to a switch in emergent patterns and produced the variable aerial patterns observed in starling flocks. Birds appear to interact with six or seven neighbors irrespective of the distance between them. Previous models had assumed the interaction to occur with all birds within a fixed distance. This apparently small change to a model led to significant changes in behavior.

Since human collective behavior also depends upon individual interactions, the StarFlag investigators are exploring the possibility of exporting models and techniques developed from studying flocking and schooling behavior to the understanding of collective economic choices—so-called "socioeconomic herding." This may lead to methods for taming excessive market fluctuations or at the very least to an understanding of how these fluctuations arise from the human perspective.

Cooperative Distributed "Intelligence"

The domination of so-called social insects—ants, honeybees, wasps, and termites—is no doubt due to their cooperative behavior, which has produced colonies that are among the most socially advanced non-human organisms in the world. Social insects are the most abundant of land-dwelling arthropods.

To gain an insight into the world of social insects, we shall look at a honeybee colony. Differences exist between them and the other groups, but patterns emerge that are inspiring and informing various technological developments of the future.

THE HONEYBEE COLONY

To many people, a bee colony is ruled by the queen bee. Wrong! The queen is an egg-laying machine. It is true that she releases a chemical messenger, a pheromone, that tells the colony that all is well with her, but as the colony expands this may become diluted and trigger the swarming instinct to split the colony. A colony is like an integrated and independent being with its own emergent group intelligence. "Emergence" is the name given to the development of complex systems from a multiplicity of relatively simple interactions. Interestingly, Darwin found honeybee colonies to be at odds with his theory of evolution in his book *On the Origin of Species*, due to the unusual reproductive structure of the colony (as described below), but eventually he realized that the colony could be thought of as a single unit from the perspective of natural selection. Individual bees do not compete against each other, and this is true of other social insect colonies, hence the description of a colony as a single integrated being. But what does this mean? Let us look inside the hive and uncover some of the apparently complex and yet, in principle, simple organization.

◀ Worker ants form living chains to bridge gaps in a trail to transport food and teammates across to the other side.

◀ Honeybees build and expand into spaces within the hive above the comb frames.

▼ A queen bee with attendant worker bees. She is about to lay an egg in an empty honeycomb cell.

SWARM INTELLIGENCE

In essence, no single bee has control over the colony or even has an overview of the colony as a whole. No central brain exists, but the colony has a distributed sensor network coupled to a sophisticated communication system and a network of integrated feedback control systems. These networks produce consensus decisions through what has become known as a type of swarm intelligence. To appreciate how this works from the bottom up, we must delve into the structure and organization of the hive.

Within the hive, the wax comb provides spaces for rearing offspring and storing food supplies, and it also acts as a communication system. It is an integral part of the superorganism. Indeed, honeybees spend most of their time in or on the comb, and even forager bees will spend more than 90 percent of their lives there. The comb is a remarkable structure built from wax secreted from young bees. The hexagonal structure is created in total darkness and hangs vertically, with the distance between combs being maintained accurately to the space in which bees can pass one another back to back without any obstruction (typically 0.3 to 0.4 inches/8–10 mm). The thickness of the comb cell walls is accurately maintained at 0.003 inches (0.08 mm), with the angle between cells walls at 120 degrees and the floor of each cell sloping gently downward to the cell base. The comb is a remarkable structure and one that has inspired engineers and mathematicians for many years.

Most animals pair (sometimes for life) to raise offspring, which in turn propagate, so successive generations are raised. Honeybees, together with some other social insects, are different. At the start of the "season" (early in the spring) the colony is made up of sterile female sisters (or workers, developed from fertilized eggs) and the fertile queen. As spring progresses, the number of workers expands and the level of activity increases as foraging begins in earnest.

The queen lays a single fertilized egg in each comb cell of the brood nest. This hatches into a larva that grows and eventually pupates in the cell, which has been capped with wax by the workers. When the new worker bee emerges, she passes through different stages of activity such as cleaning cells, wax building, nursing young, and guarding the hive entrance from intruders. When sufficiently senior, she will become a forager and leave the nest for periods to collect nectar and pollen.

Rejuvenating the Colony

At the onset of summer, worker bees build slightly larger cells into which the queen lays unfertilized eggs. These produce the larger, male honeybees (or drones). At this point, the queen and drones are essentially the reproductive organs of the colony, while worker bees in effect maintain the "pulse."

Drones have a good if relatively short life, doing no real work for the hive. They are there to mate with any young queen honeybee (a princess) that is raised by the colony. This occurs when the colony senses the need to divide, in much the same way as a gardener divides a plant that has grown too large for the site that it currently occupies. One trigger of this behavior is the reduction in pheromone per honeybee delivered by the queen, as the colony increases in size when times are good and food is plentiful. The workers build a few special cells that hang like a pendulum on the face of the honeycomb. Into each of these an egg is laid, which is then fed with so-called royal jelly, a super-rich concoction that leads to the birth of a future queen honeybee. If there is a heaven for a honeybee larva, then this must surely be it—to lie in such a rich environment. Once hatched, the new princess is likely to destroy the other developing princesses, to remove the competition. When she is approximately one week old, the virgin queen leaves the hive with a few accompanying workers for her "nuptial" flight. This is usually just one flight, but may be more, and she returns when mating has been successful.

THE BIRDS AND THE BEES

In most animal populations, a few males are able to fertilize all the females. Indeed, as we see in herds of deer, prides of lions and other large cats, groups of primates, and so on, the dominant male is the one to mate with the females, thus ensuring the "survival of the fittest" gene pool.

▶ **Drones are larger than workers and hatch from larger cells in the honeycomb. Here a drone meets the queen.**

◀ A swarm of honeybees rests in a pendulum formation, awaiting instructions from scout bees with directions to their new home.

Since all occupants of the hive are offspring of the same female, it is essential to have a method of ensuring diversity and variation of genetic characteristics, and so honeybees use the opposite strategy. For the very few virgin queens that are raised by a colony, there could be between 5,000 and 20,000 drones. How, then, can a male achieve dominance, when the potential for competition is enormous? Much of what is known about the mechanisms is still evolving—it is very difficult in practice to observe behavior.

The drones leave the hive in late morning and assemble in locations that are used each year. When out of the hive, the virgin queen uses a pheromone lure to attract the drones. This is not released until the mating flight and, indeed, the princess lives in the hive alongside drones without the mating urge being switched on. The virgin queen will mate in flight with several drones to ensure that a genetically rich pool of sperm is available for her future offspring. Each successful drone loses his reproductive organs during mating and dies immediately afterward. The young queen then returns to the hive when her sperm bank is full. This sperm will be kept fresh throughout her life and fertilizes around 200,000 eggs each year, a staggering statistic. She is unlikely to leave the hive again until triggered by the need to swarm the next year or the year after.

In the dominant male scenario, the drones would have to fight for supremacy, but instead honeybee colonies have large numbers of nonaggressive drones available at the right time of year. They pose no threat and are able to enter the hives of other colonies in the vicinity, unlike worker bees. The bad news for the drones, however, is that later in the year they are expelled from the hive and left to die because they would be an unnecessary drain on scarce resources during the winter months.

◀ The dominant male in most animal populations, such as deer, is the one to pass on his genes to the next generation. Remaining dominant, though, means continual battles with up-and-coming young males. This is not the case in bee colonies, where males have little importance and are kept merely for reproduction.

Studying the Behavior of Social Insects

Engineers and computer scientists have realized that significant technological and organizational breakthroughs are possible from studying mechanisms of social insect cooperation and applying the principles to areas such as planning, manufacture, algorithm development, communications, and robotics.

Social insects use a sophisticated communication and control system to allocate tasks to their workforce, to maintain a balance between collection and processing of nectar, pollen, and water. The principles of the foraging strategies used have inspired researchers to develop similar approaches to routing packets of information in telecommunications networks. A typical problem is establishing where communications traffic should be routed when the operating environment changes due to, say, a router being out of action. This has to take place seamlessly in the midst of dynamically changing demands.

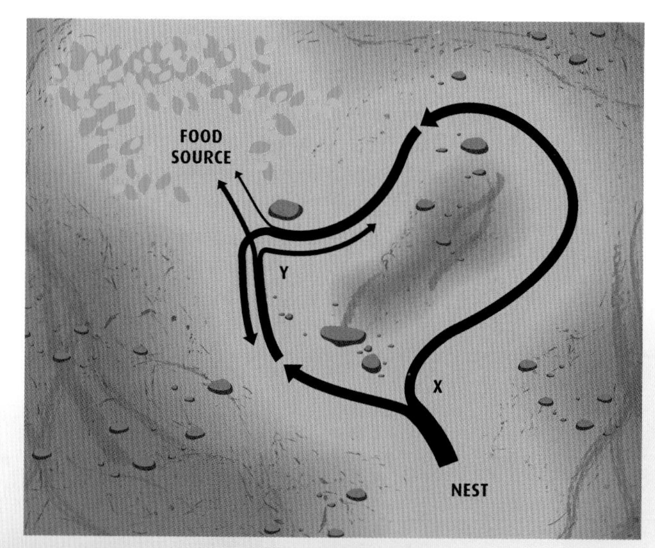

▼ A team of leaf-cutter ants on a foraging trail working together to carry sections of leaf back to the nest.

▲ Foraging strategies employed by ant colonies use pheromone intensity to label the shortest path from food source to nest and back.

Observations of ant colonies have illuminated how ants arrive collectively at "decisions" and so converge on the best solution for a given task. In one study, a number of ants set out from the nest in search of food and arrive at a junction along the path (marked X in the figure opposite). They have no previous knowledge on which to base a decision about which route is preferable or, if both paths lead to a food source, which is the shorter route. Consequently, half of the team go one way and half go the other. As they travel, each ant lays a chemical signal, or pheromone, as a message to others and to mark the way home. When the ants reach a second intersection, again about half carry on and the other half turn back along the other path that takes them to the nest. Meanwhile those on the right-hand path approach junction Y, passing the returning ants. Again they split into those going ahead to the food source and those returning to the nest via junction X. The pheromone intensity on this path back to the nest is much greater than that going forward, however, so the majority turn back to the nest, meeting the returning ants that were on the longer path. This appears, at first, a poor strategy, but now we can see that because more ants have walked the shorter branch between X and Y, from now on the ants will be more inclined to follow that route due to the dominance of the pheromone trail. In parallel with this, the strength of pheromone on the longer path decreases with time due to the slow evaporation that takes place, so eventually this path is filtered out. This is what engineers would call a "forgetting" factor in an optimization algorithm.

WHAT'S THE QUICKEST WAY THERE?

This simple method has already been the inspiration to designers of routing and planning systems. A typical "routing" problem would be the so-called "traveling salesman" problem, or the scheduling of a team of delivery vehicles, and these have been shown to benefit from the ant foraging strategy. Typical problems to be solved here might be "What is the shortest route connecting customers together?" or "How can the number of delivery vehicles be minimized?" given constraints such as capacity per vehicle, the need to start and end at the depot, and maximum distance per vehicle. Algorithms based upon the ant trail formation have been developed and shown to be a promising way of evaluating the best routes to take.

Honeybees have also developed foraging strategies and disseminate the results through communication in the darkness of the hive. A scout bee that has found a food source has to tell the workforce what kind of flower it is, in what direction it lies, and how far it is from the hive. To do this, it uses the so-called "waggle dance." The bee codes information through the angle of the main dance line to give flying direction relative to the sun, while the number of waggles or turns indicates distance. Vibrations on the comb and scent may also give clues to her sisters as to the whereabouts of the food source. Since many food sources may be available, the colony then allocates the foragers between the different sources in order to maximize return.

▶ A scout honeybee performs a "waggle dance," using angles and size of movements to give directional and distance information about a food source to sister bees.

Spreading the Load

Honeybee foraging behavior has already stimulated research into solutions for load balancing on Internet server systems, where traffic and demand cannot be known in advance. Optimization of server allocation is, however, a difficult problem to resolve due to the unpredictable behavior seen in the arrival of requests.

Honeybees must also maximize nectar and pollen flows from an array of different sources. By using the honeybee foraging strategy as a model, Internet host designers have shown that it is possible to increase revenues by 4–20 percent by using a bee-inspired website advertisement system that highlights locations that are in demand. These, in turn, tend to last longer if profitable. The advertisement "board" is effectively the analog of the waggle dance in that it communicates the location of very active websites so that other servers can be attracted within the hosting center, in order to help out with the load and to satisfy requests. The researchers are now looking to extend the analogy with the honeybee colony by making efficiency savings in energy expenditure in the same way that bees will reduce their foraging when the weather is poor. This provides a rest period for the main body of foragers, which can then take full advantage of an improvement in climatic conditions.

Important recommendations have been made on identifying the suitability of problems to a bioinspired approach. In particular, it was stressed that similarities that must exist between the technical challenge and the problems faced by the insects. This may sound rather an obvious statement, but in fact it has not always been observed. The success or failure of using an insect foraging or behavioral model to solve a technical challenge relies on the degree of overlap in problem specification. This is where definitions can prove confusing in that a bioinspired approach aims to learn from the basic principles of the biological system, whereas a biomimetic approach tends to infer copying or mimicking. Nevertheless, these recommendations are extremely important in order to ensure that the problem definition is clear at the outset.

NETWORKING

The distributed manner by which a colony of social insects is maintained without any central coordination has also inspired the development of distributed sensing and monitoring systems that can be used in a wide variety of applications. A wireless network that can effectively "mesh" nodes together when required can overcome the limitations of robustness and scalability that are associated with wireless systems. Such systems frequently require mains power for operation and are hence limited in where they can be located. FlatMesh overcomes these limitations by having every node communicate equally with every other node in the network through embedded system software on each node. No hierarchy is in place and no traditional routing system is needed, which means that any installation can be simple and quick to undertake. The network operation is then self-organizing and, since the nodes are designed for low-power operation using just a small battery, an elegant power-saving strategy has been developed.

Interestingly, this strategy has similarities with the way in which fireflies pulse their impressive collaborative flashing-light displays. Each node is built around the sensor(s) of interest, and so the range of applications is extremely large. Since each sensor node can be built into a robust container, the network can be used in hazardous environments such as the North Sea. To monitor the environment around an offshore wind farm, the nodes were fixed to buoys that were floating around the site from which the conditions could be measured and an overall picture gleaned of the changing conditions. The nodes have also been used to monitor potential landslips and flooding on railroad embankments, where traditional wired systems are time-consuming to install and where reliability can be reduced due to the cabling and connections involved. The FlatMesh system not only overcomes the installation and connection problems, but can also tolerate the loss of a node, providing that a communication pathway exists to neighboring nodes.

▼ **Balancing power generation with demand poses problems for renewable energy systems. Learning optimization strategies from insect colonies could help us to improve the balance.**

Choosing the Best Solution

Honeybees must maximize their flows of nectar and pollen when available and to do that must allocate foragers to the most productive sites. The foragers pass on nectar and pollen to collector bees at the hive entrance, which help to monitor the flow of resources and the orchestration of foragers to the best sites.

This self-regulation to maximize food collection has inspired the designers of manufacturing production systems where the problem to be solved is very similar, in that machinery settings have to be selected for a particular job and the number of options available is very large. To maximize results, and hence profits, the best combination of settings has to be chosen, and this has led to the so-called "Bees Algorithm." The algorithm is basically an optimization procedure that searches the neighborhood, but in combination with a random search. Complex, multivariable optimization problems are generally a compromise between finding "optimal solutions" and keeping to a realistic timescale. So-called "search algorithms" look for a single solution in each iterative cycle and aim to converge toward the best solution to the problem. A swarm-based algorithm, however, uses a population of solutions and so can search for a variety of possible solutions at each step. Hence, if multiple solutions exist, the algorithm then aims to locate them. Other similar algorithms include Particle Swarm Optimization (based upon flocking and schooling behavior), Ant Colony Optimization (based upon ant foraging trail formation),

▶ **Honeybee colonies have developed simple yet effective ways of allocating foragers to the most productive food sites.**

Taking insect foraging strategies as inspiration, we could orientate groups of satellite solar panels to the Sun in ways that minimize usage of fuel.

and the popular Genetic Algorithm (based upon genetic combinations and natural selection). All of these were inspired by biology. The performance of the Bees Algorithm has been compared with that of a range of other methods and shown to have remarkable robustness, particularly with regard to sensitivity to local minima and maxima. Current limitations, however, for implementation in the machine shop are the large number of tunable parameters used and the optimization of the learning mechanism. Despite being based on honeybee foraging, this will require tuning to be appropriate to the engineering application. Nevertheless, bioinspired approaches point the way forward over more "conventional" routes.

CONTROL SYSTEMS

Engineering control systems have historically been centralized because of the way in which computer power has developed from physically large systems. In parallel, sensing systems have been slow to develop, particularly sensors that can react sufficiently quickly and accurately. Theory, too, has concentrated on understanding linear systems with just a few inputs and outputs. In recent years, computing power and sensor technology have advanced dramatically and are now offering the opportunity to control systems in a more distributed and decentralized way. Theory, however, has not seen such significant strides, despite the many considerable developments underway.

Ant trail formation and optimization have inspired the development of new approaches to so-called "optimal control," which seeks to move a dynamic system such as a satellite from one state to the next in, say, the minimum time or using the minimum amount of fuel. Conventional methods of doing this have assumed that the final state is fixed. The bioinspired approach has the potential to unlock this restriction and is key to the success of controlling a group of robots or robotic vehicles where the destination point may not be defined. Central to collective robotics is the premise that cooperation can produce more than the sum of the parts.

The tasks that can be performed are well beyond the abilities of the individuals, with the added value of greater robustness to individual losses, better performance, and more efficient costs. Where have we heard this before? The social insect colony organization! Having learned from the flocking and schooling behavior of birds and fish, the collaborative team of UUVs that we met earlier now has the potential to navigate to a mission end point. The next stage is to allow the UUVs themselves to find their own mission end point. This opens up opportunities for many different operational scenarios. If the mission were to find a pollutant in the ocean or in an estuary, say, and track it to its source, the team would be deployed with an appropriate array of sensors and would have to collaborate to move together to cover the polluted area and to locate the source. Here, the mission end point would be specified by the team itself as the source is located, like the honeybee and ant foragers, which locate food sources that are unknown when they send out the scouts. Methods outlined by C. Shao and Dimitrios Hristu-Varsakelis show promise for this type of control because they are geared toward using local environmental information provided by individual team members that are required only to calculate optimal pathways to their near neighbors. Learning from insects shows us that by using distributed sensing, local knowledge, information filtering through competition, and consensus decision-making, we have the potential to achieve our goal using fairly simple rules. We could even lose a robotic vehicle or two along the way without ending the mission.

Form Follows Function

Termite colonies have impressive behavioral features that are inspiring engineers to rethink their approaches to design. Architects, too, are learning from termites. One example of where cooperative behavior can inspire building design is the humble termite mound. Is it really humble? Let us have a look inside.

▼ The airways within a termite mound produce natural airflow to maintain air quality passively in the nest. They may inform architectural design of buildings.

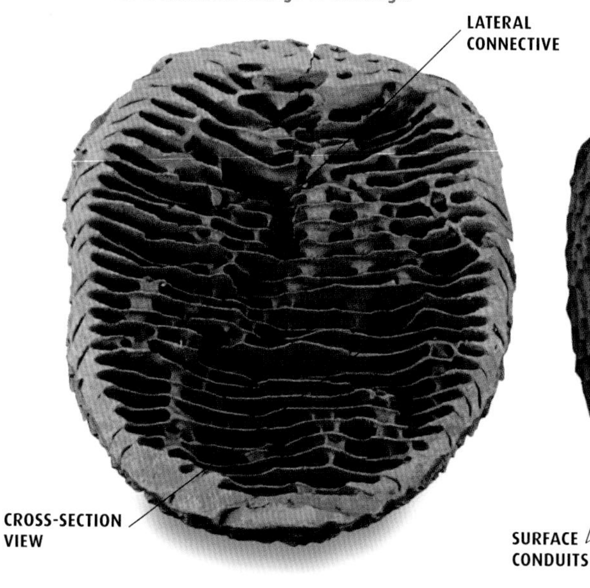

Termite mounds are really tower blocks that can stand up to 16½ feet (5 m) tall. This is impressive, given the size of a termite. In human terms, this would be a very high-rise skyscraper. Termite mounds tend to be found in rather harsh environments and have been optimized for the conditions they must withstand. Not only are the passageways and tunnels complicated, but also they have evolved to provide control of airflow to maintain air quality in the nest. Moisture levels and temperature are controlled by the airflows as well, and all without the help of an energy-wasting air-conditioning system! Perhaps a complex network of underground tunnels to emulate the termite system would allow for natural ventilation of a building, but this would require a re-think of building construction methods. Many other examples of the use of passive air-conditioning can be found in Chapter 5. In addition to the control of airflow, engineers have been inspired by the way in which a colony of termites manufactures and builds its mound. This is generating ideas for creating extremely large and rapidly manufactured structures that could be used for habitation on other planets.

▲ Building on a north-south orientation helps the termites control airflow and temperature inside the mound.

DECENTRALIZATION

Control systems are moving increasingly toward distributed and decentralized architectures as sensing and processing capabilities become smaller and cheaper. Memory has also seen dramatic increases in capability, along with reduced costs. Decentralization of control makes sense in that reliance on a single controller is removed, but problems are then created with regard to how to combine the control actions from an array of individual control systems. In addition, the physical distribution of controllers poses problems of ensuring that individual actions do not interfere with one another, but instead combine in the best way to achieve the desired response to imposed changes. In the biological world, self-regulation of a healthy organism is achieved through what is known as "homeostasis," where each "subsystem" works to maintain the balance of the whole. Resting heart rate, body temperature, respiratory rate, and so on are maintained by control loops that use feedback and interactions to operate. If one of these drops below the normal value, the control systems work to raise the value back to normal, and vice versa. In a honeybee colony, however, this self-regulation is achieved by the activities of the colony itself, and this has the added advantage that the "normal" values are adapted to changing demands. The consensus decision-making of a colony may have significant implications for the design of complex distributed decentralized systems that can adapt to changing conditions. Aircraft noise, for example, is the subject of intense research. This is the case as it applies within the cabin, as well as outside, and one significant source of noise inside an aircraft is the radiation from fuselage panel vibrations. Here, distributed control is beginning to have an impact on reducing vibrations and hence noise. The combination of distributed controllers and the placement of sensors and actuators is, however, a difficult problem, and it may be that the consensus decision-making of the honeybee colony holds clues as to how best to achieve this integration in the future.

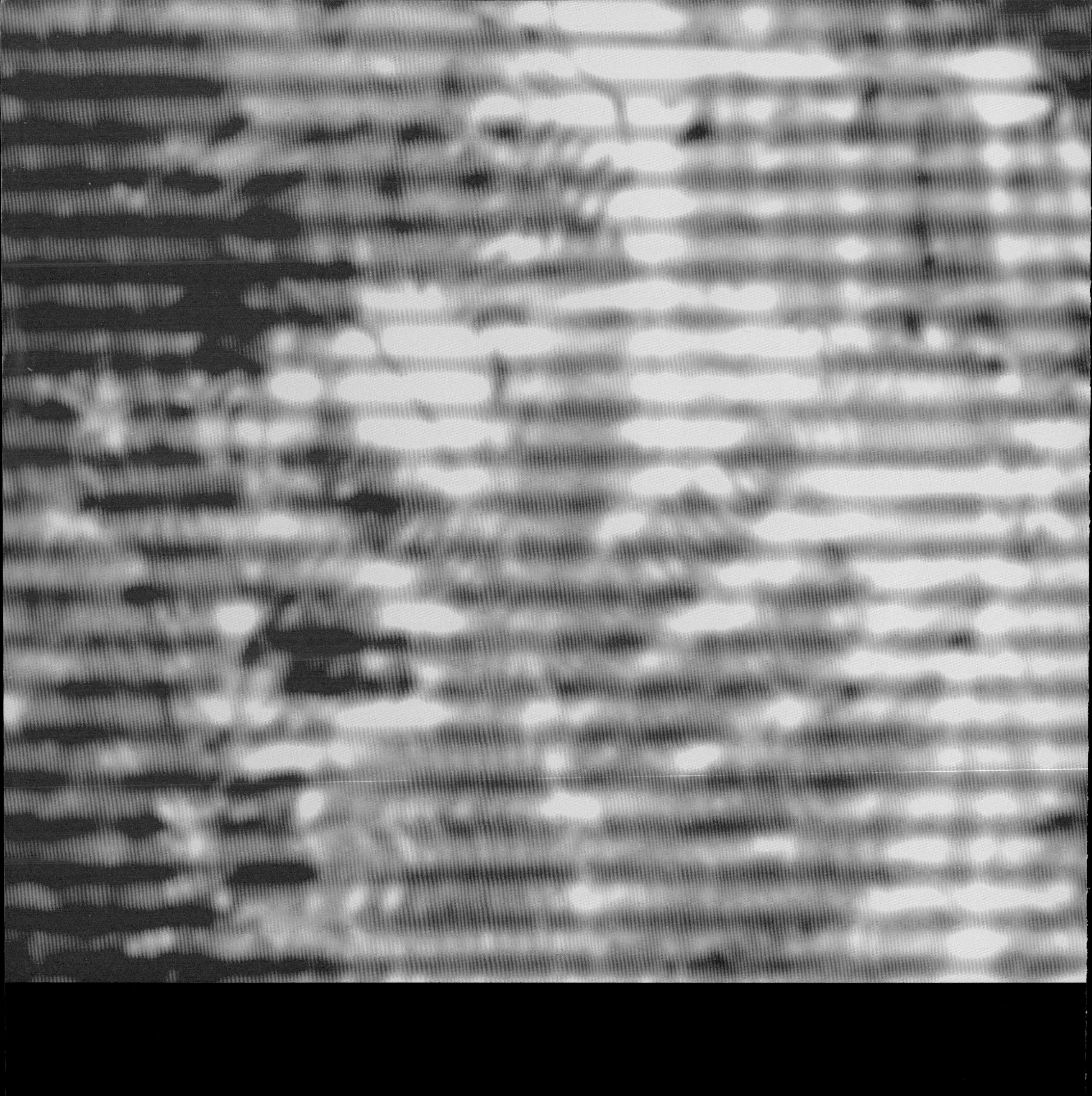

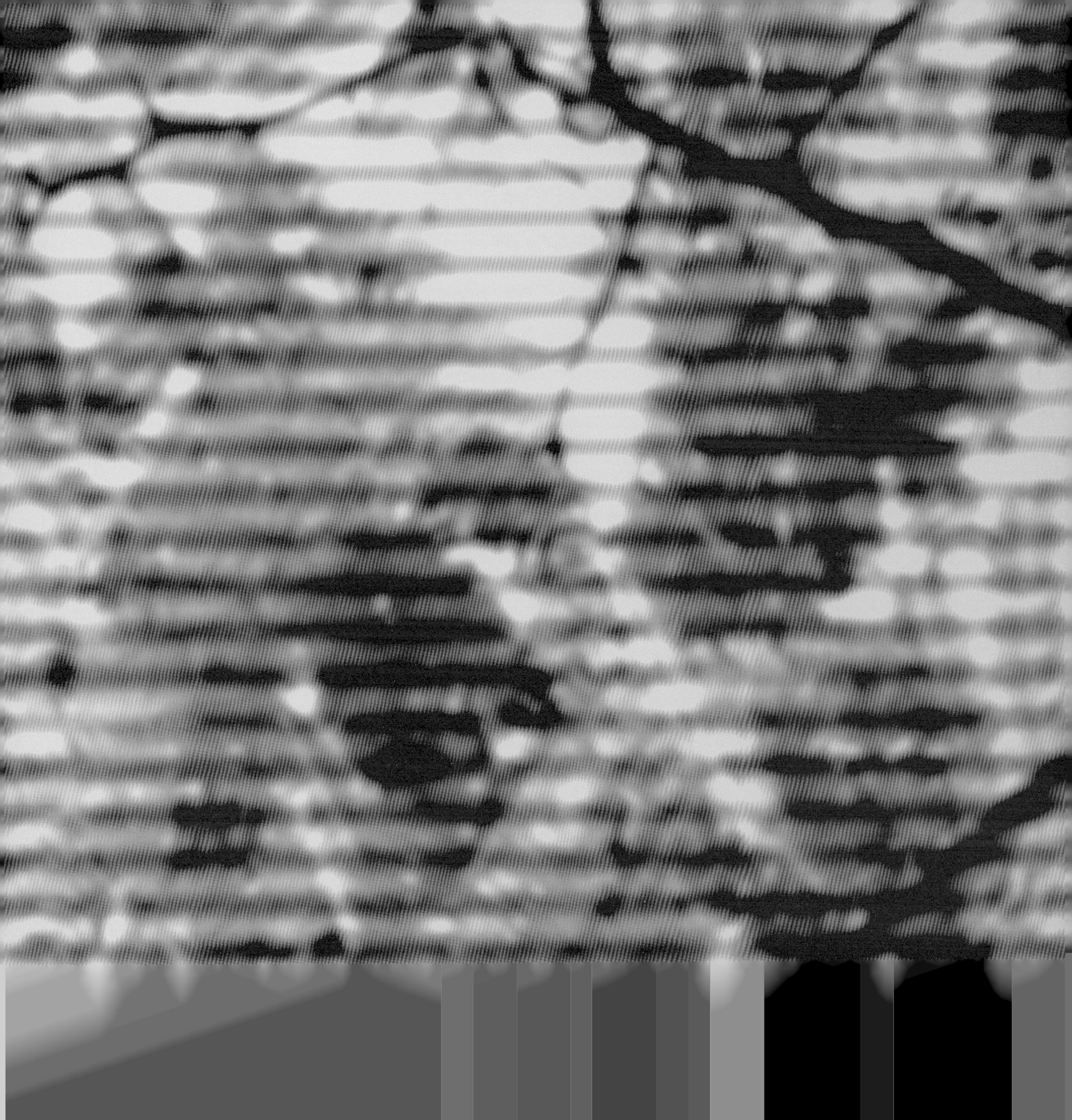

Introduction

We like the buildings we live and work in to be both well insulated and well ventilated—two ordinarily opposing processes. It would be useful to be able to capture heat from our surroundings, and also to discharge the excess heat we produce. We also need to do this while using as little energy as possible.

Early and less affluent cultures have developed cheap and simple ways to heat and ventilate structures. In the past century, however, energy has become relatively inexpensive, so our modern, high-tech culture has lost familiarity with these methods, never mind attempted to improve them. And only on the rarest of occasions have we looked for guidance at how nature does it. We have much to learn from animal and plant ventilation systems—how organisms move air and water.

Sometimes an animal's surroundings become too hot, while sometimes they get too cold. The animal might be too hot and need to discharge excess heat, while at other times it will try not to lose heat. And sometimes its other activities create thermal problems—a living creature cannot be completely isolated from its surroundings.

Animals use oxygen and have to get rid of carbon dioxide (CO_2). Land creatures need to take in and expel gas without losing too much precious water. Vaporized water carries off heat, which can be a helpful cooling method, but at the same time it leaves less water for future use. Aquatic creatures facing similar demands cannot vaporize water at all. An overheated fish is forced to do something else to remain cool. Simply pumping out warm water takes a lot of energy because water in its liquid form is very dense compared to water vapor. Average-size mammals, such as humans, use 10 percent of their energy supply merely pushing blood and air around their bodies. A fish spends far more of its energy forcing water across its gills.

◀ By moving only briefly, living in or near comfortably warm water, and accepting the temperature of its surroundings, a crocodile can avoid expending energy heating or cooling itself.

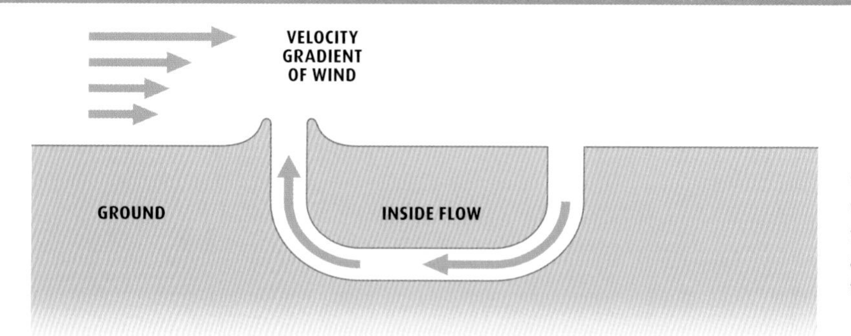

◀ Exposing one end of a burrow to stronger wind than the other will cause a unidirectional internal flow, and an elevated opening is all that is needed to expose the flow to that increased wind.

High costs drive efficiencies, in nature as well as in human invention. Organisms have evolved a host of devices to push air and water around their bodies in ways that balance any advantages with the problems also created. What follows is a description of various devices with which animals and plants move air and water around with little energy expenditure and without losing heat as fluid flows into and out of their bodies. What these represent is a set of natural devices that could be emulated by human technology.

UTILIZING OUTSIDE FLOW

Oddly enough, wherever any fluid flows across a solid, no flow occurs right at the surface of the solid itself. That is why a single swipe of a cloth cleans a plate so much faster than just running water across it. If there is no flow at the surface, then there has to be a region of changing flow speed just outward from that surface—a "velocity gradient" or "boundary layer" are the conventional terms. Within that region, flow speed increases from zero at the surface and eventually gets up to the speed of the undisturbed flow.

A velocity gradient is a kind of potential gradient, like the ones between two poles of a battery or the differing heights of water across a dam. So with a suitable converter, a velocity gradient can be a source of energy. In human technology, the most familiar version is a windmill.

A simpler converter uses a wind or water current to make air or water move through a structure. It consists of an exit opening—essentially a chimney—well above the surface and at a right angle to the outside flow, and an entrance opening (or openings) at or near the surface. So air or water enters at surface level, circulates, then leaves through the elevated exit opening.

NATURAL EQUIVALENTS

Since these systems were first described in 1972, this system has been found at work in nature. For instance, prairie dogs (*Cynomys ludovicianus*) use it to make air flow through their burrows. These are especially deep two-ended passageways. One end opens in the middle of a sharp-edged, craterlike mound; the other in a low, rounded mound. It was originally assumed the reason prairie dogs need flow through their burrows was for them to breathe, but it now seems more likely to be to track the odors of the outside world—in effect, a giant nasal system.

▶ Prairie dogs live in long, deep two-ended burrows. The air flowing through the burrows appears to carry information about what is happening outside.

Keyhole limpets (*Diodora* spp.) obtain oxygen from the water around them by using the same basic setup. Water goes in around the margins of their shells, across their gills, and is drawn out of the openings—the keyholes—on the upper surface. Some giant termite mounds (of *Macrotermes michaelseni*) draw in air through openings or porous areas around their bases, supplying oxygen and eliminating carbon dioxide, to sustain perhaps 44 pounds (20 kg) of termites. Some sand dollars (*Mellita quinquiesperforata*) draw water from sands beneath them up through slots running outward from near their centers, giving the lower surface and the mouth better access to tiny edible creatures living between sand grains.

INPUT AND OUTPUT

Central to each case is a connection between input and output through which fluid can flow with little resistance. It amounts to a pump driven without any direct costs, but one that produces impressively high flow rates with only a very little pressure. On the one hand, the scheme offers full independence of the direction of wind or water current, a distinct advantage with uncertain winds and tidal currents—it is an omnidirectional windmill. On the other hand, the internal flow depends disproportionately on the outside flow, so these devices are awkwardly dependent on the speeds of winds and water flows.

CONVERTERS IN THE HUMAN WORLD

The fire in an open hearth roars when the chimney above it feels a gust of air. In short, we find this device unremarkably familiar. Ventilatory cupolas that work in the same way have long graced a variety of buildings, from schoolhouses in open country to storage facilities for hay, grain, and other crops—places where airflow provides increased comfort and decreased humidity. Preindustrial societies built (and still build) many structures with openings at the top for ventilation. The yurts of Central Asia commonly have a porous felt top, and the yarangas of Siberia look as if they have a similar arrangement. Inuit igloos often have small ventilatory holes near or at their tops. And all the lavvus of the Sami of Scandinavia and the similar tipis of the plains Indians of North America were open at the top. Within structures of this kind, open fires could be kept going without asphyxiating the occupants. These lodgings usually had some way to cover the openings at their tops when necessary.

USING UNIDIRECTIONAL FLOWS FOR VENTILATION

When wind fills a sail broadside to it and pushes a boat leeward, it causes a dynamic pressure. That same pressure can drive flows of either air or water. A dynamic-pressure system will usually exceed what an internal–external flow could produce. At the same time, however, to position and orient input and output openings correctly, you have to know from where the flow will come.

▶ Native American tipis have an opening near the top that faces downwind in the steady breezes of the prairies. This opening draws smoke out of the tipi, which permits an open fire to be lit in the center of the floor.

◀ Humans have long used the dynamic pressure of the wind to propel boats across water. We have figured out both how to keep from being blown over and how to sail upwind.

▼ Water goes in the shark's mouth and out the vertical gill slits farther back, driven by that same dynamic pressure and the swimming of the animal. Some sharks have to swim constantly or they will asphyxiate.

FUNNEL FEEDERS

Organisms can count on unidirectional flows in at least two circumstances. First, nontidal rivers flow reliably downhill, so any creature living on the bottom can depend on that dynamic pressure. Some insect larvae build U-shaped burrows of silk and pebbles. The upstream end of a burrow ends in a flaring trumpetlike input, while its downstream end is flush (or nearly flush) with the riverbed. A silk net spun across a wide part of the funnel separates the edible tidbits from the passing water.

RAMJET SYSTEMS

The second situation where one sees unidirectional flow is where the animal itself moves. The faster something swims or flies, the greater the pressure will be on its front end, and again pressure increases disproportionately with speed. That makes the whole scheme a nearly perfect driver for respiration because faster swimming or flying demands the same kind of disproportionate increase in the consumption rate of oxygen.

The classic example of this is what is called "ram ventilation" in fish. The water flows in through the mouth of a fish, crosses its gills, and leaves behind the gills. For fish such as salmon, ram ventilation works in concert with the water the fish pumps with muscle. In large fish such as tuna and certain sharks, ram ventilation does the whole job, and they have to swim to breathe, just as they breathe to swim. Some large insects drive air passing their flight muscles during flight through the passageways in much the same way.

Harnessing Multidirectional Flow

Dealing with flows that periodically reverse—such as tidal zones—takes additional equipment. The tunicate Styela montereyensis *filters food from seawater. It attaches itself at one end to a rock and allows the other end to reorient like a weathervane. Nothing fancy—just using passing flow to bend it in the stream's direction.*

This method helps the tunicate to keep its input facing upstream and its output at a right angle to the flow. Sponges have a less obvious way of managing shifting flows. Flow through sponges combines pumping and using local currents. The sponges pump water actively at the same time as utilizing currents—pumping at a remarkable rate of around a body volume every five seconds. How can sponges keep their downstream openings from undoing the benefit of their upstream ones? Tiny one-way valves on each pore ensure that the pumping and filtering chambers are fed only by the pores that experience inward pressure.

HUMAN APPLICATIONS

People use the same arrangement in all sorts of applications, but it is ordinarily only a minor element. Dwellings in warm climates are often oriented so a large portal faces upwind. Some car engines obtain air at a slightly increased pressure by directing their input opening forward, taking advantage of the wind their vehicle's motion causes. Of greater significance, modern cars turn off their cooling fans when they travel fast enough (which is not very fast) to provide sufficient airflow through their radiators. Their setup differs little from the opportunistic ram ventilation of fish, such as salmon. Aircraft, operating at higher speeds, make

▶ The fan of a fanjet engine forces air both into and around its combustion chamber. It gets help from the oncoming wind caused by the aircraft's forward motion.

greater use of dynamic pressure. This type of pressure plays a substantial role in modern jet engines, although the extreme case, the ramjet engine, is not in widespread use.

The principal limit on driving flows with oncoming air comes from the relatively low pressures available at all but very high speeds—especially when the moving fluid is air, rather than water. Only applications that need their power in the form of high flow rates and tiny pressures can benefit. Supercharging an automobile engine with an air scoop yields little benefit at any road speed. Similarly, a flying bird cannot reduce the work of breathing by opening its beak as it flies—at a speed of 65 feet (20 m) per second, the pressure gain would be only a few thousandths of an atmosphere.

DOUBLE-ACTION VENTILATORS

Many of the creatures that use an external current to drive some interval flow combine the pressure drop at output openings at right angles to the flow, to create a pressure boost at the upstream inputs. An abalone (*Haliotis* spp.), a mollusk that looks slightly like one half of a clam, has a row of holes along its upper surface. The upstream holes face into the flow and act as inputs, while ones farther along the shell open at right angles to the flow or slightly downstream, and provide outputs.

A ram-ventilating fish uses the same trick. Not only does its mouth face upstream as it swims, but also its outputs on either side are located near the widest part of its body. With that arrangement it can take full advantage of maximum pressure on its input and minimum pressure at the output.

PUSHING AND PULLING WATER

Low pressure at the outputs combines with high pressure at the inputs to create a force that can be harnessed to save energy. Consider a swimming scallop. It claps its shells open and closed to move in short hops. The gaping "mouth" faces forward, while the hinge is at the rear. A pair of jets, one on each side of the hinge, provide propulsion. The squirt of these jets comes from closure of the shells and that, in turn, comes from contraction of the large adductor muscle between them—that's the bit we eat. The scallop's shell is not opened by a muscle, but rather by elastic pads next to the hinge. Once in motion, the open stroke is assisted by the combined positive pressure during the open gape and the reduced pressure on the sides of the shells. Similarly, baleen whales such as fin whales (*Balaenoptera physalus*), which gulp-feed as they swim, use an inward push in front and outward pull on the side (or lower side) to reopen their enormous mouths and expand their throats. Contraction of the muscles of their conspicuously corrugated throats pushes water and food through their baleen plates as the mouth closes. So as the whale swims that combination of muscle contraction followed by pressure-driven expansion pumps the vast volumes of water it needs to process.

UTILIZING WIND AND WATER FLOW

Sponges are an excellent example of how nature uses external currents to drive internal flow. They have output openings at the top at right angles to their inputs. This results in strong flows through the sponges and helps to prevent them from sucking water already filtered for food.

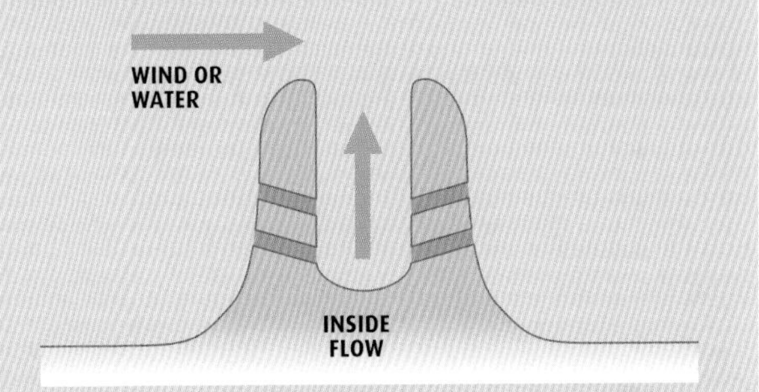

▲ Wind or water movement can induce internal flow both by increasing the pressure on the upstream openings and by decreasing the pressure at the exit hole on top.

◀ Sponges must process just over 35 cubic feet (1 m^3) of water to get a few grams of food. Their pumping is augmented by flow induced by the movement of water around themselves, as shown in the illustration above.

Staying Cool

Until recently, houses in warm, humid parts of the United States lacked air conditioning, so people paid attention to opportunities for natural ventilation by using what was available in the environment around them. Yet in many cases they were using systems already "discovered" and used by insects and plants.

In the north, houses had small windows and irregular-shaped interior hallways. In the southeast, a common design combined windows that extended nearly from floor to ceiling with a central hallway that led directly from front door to rear. With even a gentle breeze, opening both doors produced a good flow of air. This drew air through open windows into the rooms, then into the central hallway. Such houses had other tricks. Their large windows were tall rather than wide, taking advantage of thermal buoyancy—rising hot air. The windows were opened at both top and bottom. Warm air left the building at the top, drawing in cooler air from outside. Porches surrounded a house, blocking walls and windows from direct sunlight. Finally, walls and porches were shaded by trees, rather than being surrounded by wind-blocking low shrubbery.

CONCENTRATING WIND

Double-ventilation devices can form the basis of what might be called "wind concentrators." Ordinary wind turbines have very long blades that turn slowly; they are heavy and need to be mounted far above the ground, with a substantial speed-increasing gearbox. Alternatively, one can envision an erect conical or cylindrical structure with skyward upper opening and openings around the sides. The side inlets would be near the ground, or arranged (as in sea sponges) so that they are open only when facing the wind. Since wind could then blow a lot faster inside than outside, these wind concentrators could replace the big, high rotor with a small turbine that has a rapidly turning vertical shaft.

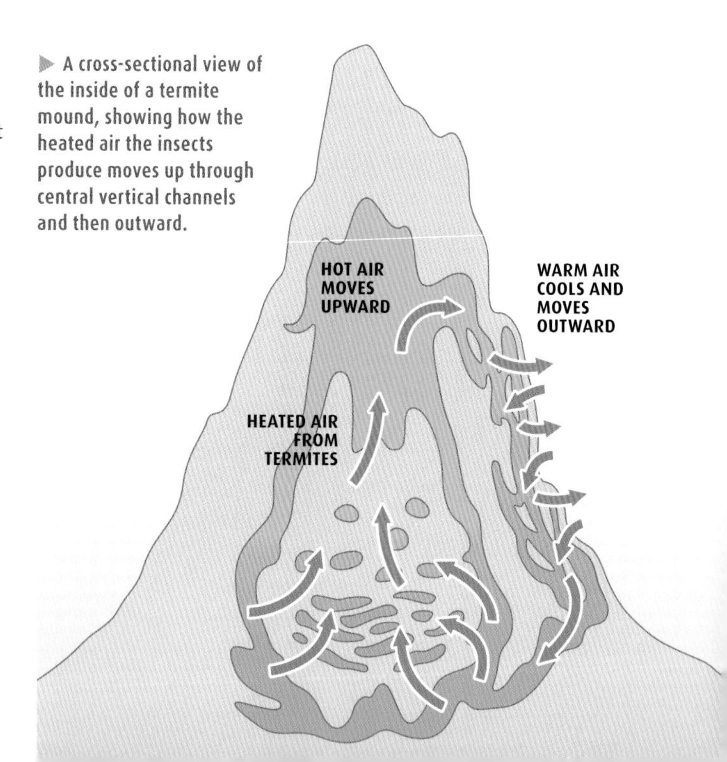

▶ A cross-sectional view of the inside of a termite mound, showing how the heated air the insects produce moves up through central vertical channels and then outward.

◀ Thermal buoyancy, caused by termites using oxygen and producing carbon dioxide, drives ventilation through their tall termite mounds, and helps to cool the rising hot air. A similar process is at play in skyscrapers filled with moving people.

DRIVING VENTILATION WITH RISING HOT AIR

A chimney draws even when the wind fails to blow. Gravity provides the driving force. Heated air rises because it has a lower density than cool air—as cool air sinks, heated air floats upward—so-called "thermal buoyancy." The giant mounds of some African termites, in particular those of *Macrotermes natalensis*, show how thermal buoyancy can drive ventilation. Thousands of termites use a lot of oxygen and produce a lot of carbon dioxide. At the same time, they generate lots of heat. The soil used to make the mound does not conduct well. The mounds are therefore arranged so that the heated air they produce moves upward through central vertical channels. It then moves outward (still inside) and, as it gets cooler and denser, moves downward through another set of channels just beneath the surface of the mound.

CHANNELING UPDRAFTS

Where else might we run across large, immobile terrestrial organisms that have problems with excessive heating? All that thermally driven ventilation demands is that the system's heat is distributed unevenly in space or time. Possible candidates are the largest cacti, particularly the saguaro (*Carnegiea gigantea*) and barrel (*Ferocactus acanthodes*). All these massive plants have large, vertically oriented corrugations on their surfaces. Spines are located outside the corrugations and therefore should not impede upward flow through these channels. Corrugations let the plants swell and shrink as their water content changes. But the channels might also facilitate passage of thermally generated updrafts that flow close enough to the surface to prevent strong sunlight from overheating the plant.

FREE CONVECTION IN LEAVES

Even structures of modest size can make use of thermal buoyancy using what is called "free convection." The thin leaves of ordinary trees never feel very warm, even when in still air and full sunlight, because of the greater thermal mass of fingers than that of leaves. Your fingers draw heat from the leaf; as you touch it, you cool the leaf more than it heats you. Broad, thin leaves can reach 36° F (2° C) above air temperature, and that risks damaging the enzymes in the leaf's cells. While in most places air is rarely still enough for a temperature increase of that magnitude, the leaves' low masses relative to their areas make them heat quickly in the briefest of lulls—typically a degree every few seconds.

At airspeeds below what we can feel—8 inches (20 cm) per second—free convection transfers more heat to the surrounding air than does wind-driven, or forced, convection. The lobed shapes of many common leaves, such as most species of oak, improve the coupling between the leaf surface and air to facilitate free convection. Unlobed leaves also have another instructive tactic. They avoid being horizontal, which gives the poorest convective cooling, or they wilt downward toward a safer, vertical position when the sun is bright, the air is hot and still, and water is scarce.

Using Thermosiphoning

In a closed system, buoyant fluid will rise only if an equal mass of cooler fluid falls. Circulation from this natural convection is known as a "thermosiphon." Our technology makes more use of thermosiphoning than nature does. In one guise or another, we depend on it to cool much of our electronic circuitry.

This kind of ventilation is completely dependent on air and gravity, so it will not work either in a vacuum or in the weightlessness of spacecraft. One hears claims that the design of buildings that take advantage of thermosiphoning, driving flow either with the heat of people and appliances, or with solar heating, is biomimetically derived from African termite mounds. The basic scheme, however, has been clearly recognized at least back to the work of British scientist Jean-Théophile Desaguliers (1683–1744) in 18th-century England.

We occasionally put double roofs on houses, with air passageways between them and a vent along the top ridge. Sunlight heating the outer roof generates convective air currents that keep the inner one from getting much above air temperature. We more often put insulation in the floor than in the ceiling of an attic, while providing openings in through the eaves and out through a ridge vent. The outer layer of a double roof does not need to be particularly sturdy, and tents of fabric, sometimes covering several dwellings at once, have occasionally been used.

REDUCING SOLAR HEATING

Warming oneself by direct exposure to sunlight makes good sense in cold weather. It is less effective, however, when air temperatures are high—or when you are generating a lot of internal heat. Properly managing your solar input can reduce your need for ventilation, ultimately reducing it to no more than what is required for respiratory (or, for plants, photosynthetic) gas exchange. We humans shed clothing to reduce our insulation, while retaining garments that limit insolation—exposure to sunlight. Inhabitants of deserts have never espoused nudity as the best way of keeping cool.

▲ A desert lizard adjusts its posture to face the sun directly and thus expose it to as little sunlight as possible—noticeable in the small shadow it casts.

▲ These leaves intercept most of the light that falls on them when, as here, they are in the shade. In direct sunlight, each tiny leaflet rotates lengthwise to let the sun's rays pass through the leaf.

REFLECTIVE POSE

Other organisms that are unable to seek shade control their solar exposure using two mechanisms. They may assume postures chosen so that they intercept as little direct illumination as possible, something evident at a glance by observing their very small shadows. Or they may coat themselves with materials that absorb little light or heat.

Controlling body temperature by postural adjustments is usually described as "behavioral thermoregulation." Terrestrial insects and reptiles provide the best examples. On a cold day, a moderate- or large-size insect will adopt a position that intercepts as much sunlight as possible, while staying close to the ground to limit wind-driven cooling. The same insect may stand erect in the heat of day, often with its head pointed downward and its abdomen upward, so that it casts a minimal shadow. Lizards and snakes bask in sunlight to gain heat. When too hot, they often minimize input by burrowing into the ground or seeking shade.

SHIFTING PLANTS

Many plants adjust their positions to control their exposure to sunlight. As mentioned earlier, leaves wilt downward to improve ventilation by free convection. Such changes of orientation will also, however, reduce the incidence of sunlight on their surfaces—plants need the energy in sunlight to build sugars from water and carbon dioxide.

The silk tree (*Albizia julibrissin*), often planted as an ornamental, takes the game a step further. Its leaves, with their numerous leaflets, have three distinct postures. When a leaf is shaded, it exposes the tiny leaflets to the sky. If a leaf in this configuration is moved into direct sunlight it casts a large shadow, showing how well it intercepts light. At night, the leaflets of sky-exposed leaves collapse together against the leaf's central rib, minimizing exposure to the cold sky. In full sunlight, individual leaflets rotate lengthwise, intercepting as little of that direct light as possible, and thus casting almost no shadow.

CASTING SHADOWS

Some buildings have long used an equivalent device. As in nature, what matters is minimizing the exposure to direct sunlight on surfaces that absorb and conduct heat. Awnings and eaves are perhaps the simplest structures. Well-proportioned eaves shadow walls when the sun is high, but also allow direct illumination when the sun is nearer the horizon—in the cool of dawn and dusk, and in winter.

Dwellings are not the only human structures that might benefit from less direct solar exposure. Cars in parking lots could also be shaded by nearby trees. Parked cars would be cooler in summer, with less energy required to cool them after a long period of being closed up. Urban heating would be reduced by increasing the overall reflectivity of cities, thus lowering the costs of air-conditioning. One might do the same for roads and highways as well. Paved surfaces now represent a major proportion of urban land area, so the effects of shading them would be significant.

Tuning Solar Input

A perfect reflector receives no energy from radiation striking it and has to reflect the specific kind of arriving radiation, whether visible light, infrared, or ultraviolet. And while silvery metallic lining for coats was once sold for its "insulating value," long-wave infrared radiation, indifferent to the silvery color, passed through it.

Better physics underlies claims that white fabric absorbs less solar radiation than does black fabric. White cars stay cooler in the sun than do black cars, and houses with roofs of a light color need less air-conditioning than do ones with darker roofs.

ANIMAL COLORING

Thinking along those lines, we should expect large animals to have light-colored fur, especially those living in warm, sunlit places, and that small animals would be darker, especially in cold polar regions. Yet reality matches that expectation only loosely. The color of a coat is more likely governed by the animal's need to hide or be seen than its need to absorb or reflect heat. Camels and gazelles of Africa are not particularly white, while musk oxen and lemmings

◀ **The white color of this structure minimizes the heating effect of direct sunlight striking it, one of many passive climate-control devices used by traditional constructions.**

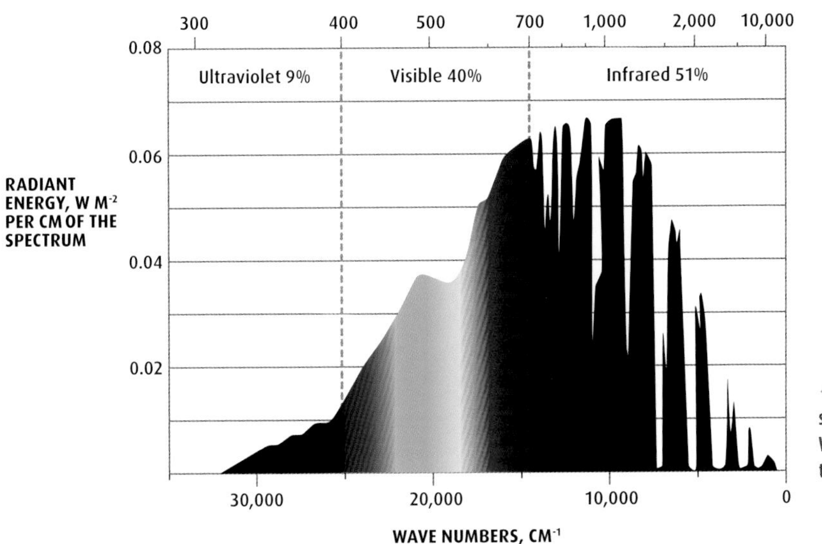

◀ More of the Sun's energy arrives at the Earth's surface as infrared radiation than as visible light. Whether a surface absorbs or reflects the infrared therefore greatly affects its solar heating.

of the far north are far from black. The discrepancy in color is a clue to another literally invisible factor that plays an important role. We most often see the spectrum of radiation hitting Earth in graphs that plot the intensity of sunlight against its wavelength. Most of the sunlight comes in the visible region. That can be misleading. The energy of radiation varies inversely with wavelength. So we need a different axis if we want to view how energy content varies over the spectrum of sunlight. That can be done by making a graph of wavelength (or wave number) against energy. In such a graph, energy within any range of wavelengths is represented by the area under that part of the graph.

INVISIBLE HEATING

This graph carries an important message. Most solar energy that reaches Earth's surface does not come as visible light. Whatever sunburns it might cause, only a negligible portion comes in the ultraviolet. The energy in the invisible infrared exceeds that in the visible region. When it comes to the heating effect of sunlight, the infrared "color" (the infrared wavelengths reflected) of a plant or animal has more impact than the organism's color when viewed in visible light.

The relevance of the infrared was first recognized in work on vegetation. Leaves turn out to be white—that is, non-absorptive—in the infrared. They stop absorbing when wavelengths get just longer than those suitable for photosynthesis. After all, absorbing this stuff would just make them hotter, risking damage to their proteins. We know less about animals, but birds' eggs reflect a great deal—typically 90 percent—of the near infrared. The shells of desert snails seem to do about the same.

ARTIFICIAL FABRICS

Most pigments, fabrics, animal skin, and fur absorb in the near infrared. Artificially colored materials are therefore often black when viewed in that region of the spectrum. Artificial foliage used to hide battlefield installations could be distinguished easily from real shrubbery using infrared cameras.

The best insulation will have to be created with the infrared in mind. A fully white roof—reflective in the near infrared as well—should be better than a merely white-looking roof at reducing the inside temperature of a building in a hot place.

Storing Heat to Reduce Extremes

Sunlight, air temperature, and wind all vary over timescales from seconds to years. Organisms often store heat to buffer themselves against the extreme fluctuations of temperature in their surroundings. The timescales over which they store heat and the diversity of organisms using storage range widely.

▲ Stone plants, much smaller than camels, manage the same heat-storage trick by coopting the surrounding soil to store far more heat than they could alone and thus slow their heating without expending water.

The classic case is the dromedary camel. The combination of hours of inescapable sunlight and the heat produced by a mammalian body could be remedied only by vaporizing a lot of precious water to keep its body temperature at acceptable levels; however, the camel's water supply is limited. As a result, the camel allows its body temperature to rise, confident that night will surely follow day, with lower air temperatures and an open, cold sky under which to cool down. Unusually for a mammal, a dromedary's body temperature rises from about 93° F (34° C) to 104° F (40° C). This thermal tolerance halves the animal's overall water loss.

◀ Camels are big creatures, so they gain and lose heat slowly. That slowness also allows them to warm slowly enough for night to arrive before they become lethally hot.

HEAT STORAGE IN PLANTS

One might expect heat-storage systems to work best for such massive organisms. At least one kind of small plant does much the same thing, however, taking advantage of the ability of soil rather than flesh to store heat, much as people do with some underground buildings. These are the so-called stone plants (*Lithops* spp.) of the deserts and scrublands of South Africa, which stick up only a few millimeters above the ground. Soil not only stores heat well, but also conducts it poorly, so the temperature of

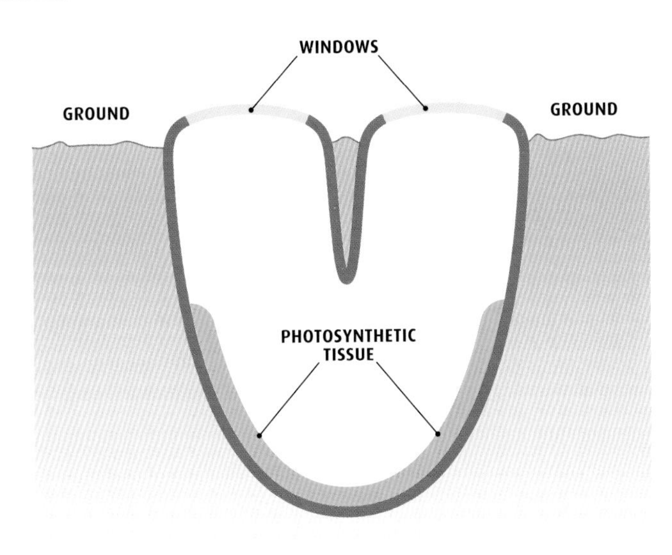

◀ This cross section shows the critical active tissue of a stone plant. This tissue is vulnerable, which is why it is located well beneath the very hot surface soil of the desert in which it lives.

▼ We use massive iron radiators to slow the transfer of heat generated by a furnace to the rooms in which we live. That way, room temperature fluctuates little as the furnace cycles on and off.

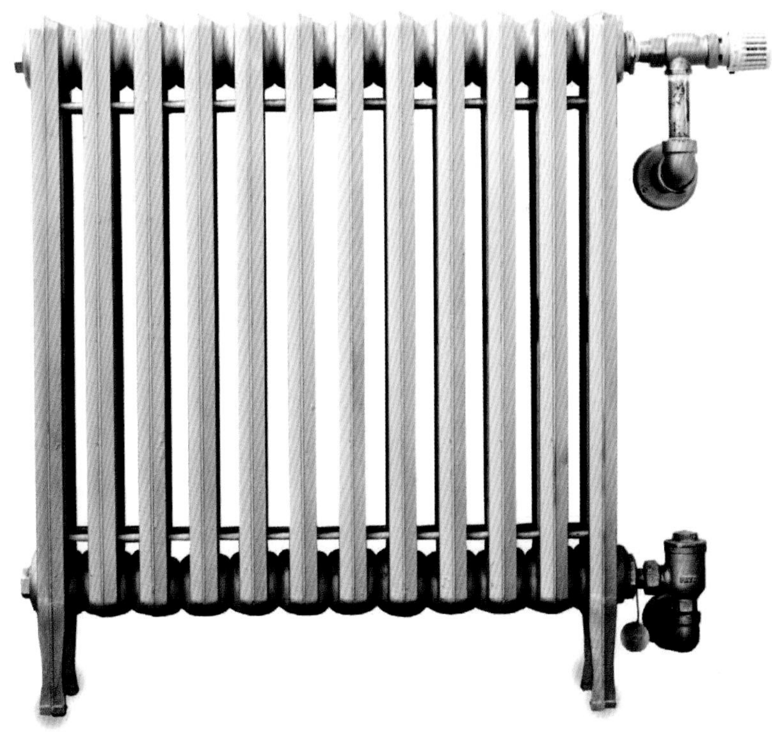

sunlit soil decreases rapidly with depth. Only the upper layers are heated. Stone plants take in light through translucent "windows" near ground level and put their photosynthetic tissue down where the soil is cooler.

In cross section a stone plant looks like an upward-gazing vertebrate eyeball. Day-to-night temperatures inside the plant vary more than a camel's body temperature does, but fluctuate far less than if the whole plant were on the surface.

HEAT STORAGE IN BUILDINGS

Any heating (or cooling) system that goes on and off depends on heat storage. All that differs are the extent and timescale of the storage. Most household heating systems cycle on and off. How rapidly they cycle depends mainly on the capacity of the heater and the house to store heat. Forced hot-air systems typically cycle more frequently than do steam-heat systems because the latter have massive iron radiators, which store a lot of heat.

Sometimes we make more explicit and deliberate use of heat storage. Some solar-heated houses pass heated air through piles of loose rocks in their cellars, making the rocks accumulate heat when the sun shines, and blowing unheated air through the rocks at night. An occasional underground building achieves year-long temperature buffering through heat storage in the surrounding earth. Most residential heat pumps use outside air both as a heat source for heating and as a heat discharger for cooling. Pipes buried in the ground yield greater efficiency, if at a greater initial cost. They carry out a season-to-season version of the trick used by camels and stone plants.

Conserving Heat with Exchangers

Once at the desired room temperature, the ideal home needs no heating system at all. Nor should it need any air-conditioning beyond what is needed to dispose of the heat from our bodies and appliances. Everything else can be attributed to wastage with heat lost due to poor sealing.

A perfectly sealed house, though, would quickly begin to smell, thanks to the waste and other products of our bodies and food. Is it possible to evade that problem, in effect, to move gases or liquids between places at different temperatures without at the same time moving heat?

Our technology is rife with heat exchangers, devices such as those ill-named radiators of water-cooled automobiles. What the ideal house needs is a heat exchanger that transfers heat from outgoing air to incoming air (or vice versa in hot weather), so that heat reenters the house, rather than being lost to the atmosphere. The most efficient way to conduct such an exchange uses what the engineers call a "counterflow exchanger" and the physiologists call a "countercurrent exchanger."

Whatever name this process is given, the passing fluids move in opposite directions. In that way, the outward flow from which an inward flow draws heat starts at room temperature. Simultaneously, the inward flow to which an outward flow loses heat starts at outside temperature. If heat transfer was total at all points along the exchanger, no heat would be transferred from inside to outside—even though fluids flowed both ways—giving ventilation with no leakage of heat.

▼ If a wading bird kept its unfeathered feet at body temperature, it would, in effect, expend most of its energy in a futile attempt to heat a lake or ocean.

COUNTERCURRENTS IN NATURE

The first specific recognition that organisms might make use of countercurrent exchangers was in a suggestion made by French physiologist Claude Bernard (1813–78). He suggested that heat might be transferred in the blood vessels of human arms from outward arterial flow to inward venous flow. Although this is true, it is a relatively ineffective heat exchanger. More and better exchangers came to light in the 1950s and later, following an insight by Norwegian–American physiologist Per Scholander (1905–80) that a long-known anatomical structure of intermingled and subdivided arteries and veins, a "rete mirabile," worked as a countercurrent exchanger. Its odd intermingling of small vessels conducting blood in opposite directions provided just the right intimate thermal contact necessary for good heat exchange.

The classic rete in nature is the one in the vessels that supply the legs of wading birds. It is located just above where the legs emerge from plumage. So equipped, a bird can in effect connect cold-blooded legs to a warm-blooded body and avoid losing energy by heating lake or ocean water. Similarly, the flukes of small whales and dolphins stay cold while supplied with blood at mammalian body temperature.

When running, gazelles and sheep keep their brains from getting as hot as the rest of their bodies by using a rete in their heads. Ascending arterial blood is cooled with venous blood that has been chilled evaporatively in nasal sinuses. Most retes can be bypassed by opening larger blood vessels running parallel, so that an animal can dissipate heat only when it produces too much, rather than too little.

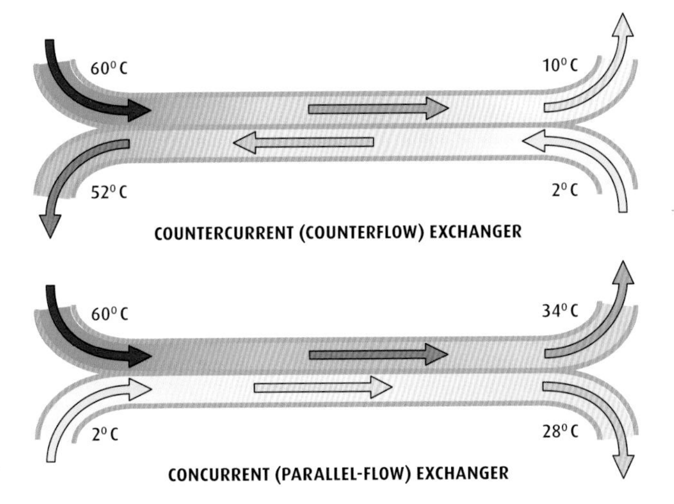

▲ A heat exchanger achieves maximal efficiency if the two fluids, whether air or water, run in opposite directions as heat moves from one set of pipes to the other.

EXCHANGER TECHNOLOGY

In his 1922 description of the crude exchanger in our arms, Scottish physiologist J. S. Haldane (1860–1936) refers to a technological analog, a regenerative exchanger on a furnace. Combustion furnaces need sources of air for their fires and flues for combustion products, and no furnace transfers all its heat to whatever it is charged with heating. An exchanger draws in outside air through a pipe in thermal contact with its flue (exit) pipe. Gas in the flue goes outward; the supply for combustion comes inward. This arrangement was certainly known in the 19th century, although one cannot always distinguish between deliberate use of the counterflow principle and some scheme designed to keep cold combustion air from causing an unpleasant draft.

As the payoff from energy efficiency has increased, heat exchangers have come into ever wider use. But most, like our automobile radiators, are less efficient "crossflow" rather than "counterflow" devices. Counterflow exchangers are oddly awkward to make. Two flows going in opposite directions must make intimate thermal contact while not actually mixing. That takes a very long pair of parallel pipes, which proves a cumbersome feature in pump design. Or it takes two sets of parallel pipes, with complex branching systems on each end of each set. Sometimes exchangers have only one set of discrete pipes, with flow in the other direction running in between them and constrained only by an outer jacket, but that just drops the complexity from a four- to a two-branched system. By contrast, organisms find production and proper arrangement of tiny pipes a simple matter, at least judging from the number of times countercurrent exchangers have evolved in different species.

Breathing While Conserving Heat

A full breath consists of inhalation and exhalation. That reciprocating flow is not only how humans breathe, but also how all terrestrial animals that pump air do it. Even some insects pump air in and out of their bodies—mainly large ones that are flying and need a lot of oxygen, as well as to discharge excess heat.

This in–out reciprocal arrangement cannot be considered ideal because it suffers from two basic problems. First, reciprocating flow means accelerating and decelerating both air and some amount of pumping machinery, which takes energy. We expend 1 to 3 percent of our entire energy production ventilating our lungs, and the percentage goes up when we are active and have the least spare capacity. Second, breathing inevitably fails to provide a full exchange of air. Respiratory "dead space" is an appropriate name for the volume of pipes that connect the outside air with the alveoli deep in the lungs, where the exchange actually takes place. In addition, the lungs have a nonbreathable residual volume. What is left is the volume, a fraction of that of the total respiratory system, that we can actually exchange with each breath.

Air-breathing vertebrates such as ourselves could possibly do better with a more direct connection between lungs and the outside, one less entangled with our digestive system. Vertebrate evolution, however, has just not gone that way.

▶ Even when fully pressed, some liquid remains in a hypodermic syringe. That volume corresponds to what we call "residual lung volume" and "dead space" in respiratory systems.

BEST OF A BAD JOB

In nature, a bad legacy often generates evasions and devices that, in effect, make a virtue of necessity. Running quadrupedal mammals minimize the work needed for that unavoidable acceleration and deceleration. Their abdominal contents hang from their backbones and swing back and forth as they run. The swinging, in turn, pumps their lungs through pistonlike movements of their diaphragms as long as they synchronize the motions of running and breathing. Kangaroos do the same as they jump, and some birds have rib cages just flexible enough that their flight muscles can squeeze them enough to pump air.

At least dead space can help to condition air. A one-way flow, however superior for gas exchange, might subject the lungs and all the blood that flows through them to excessively cold or dry air. Air that must travel through a long pipe from the outside to the alveoli can be warmed and humidified en route; however, that also costs both expensive heat and precious water.

Many mammals and birds, particularly small desert creatures, evade these problems with a remarkable version of the countercurrent exchanger, as just described. An exhalation loses heat to the surfaces inside the nasal passages, which have been cooled by the previous inhalation. The subsequent inhalation gains heat from those surfaces, which were warmed by the previous exhalation. And as a result, water condenses out on exhalation and revaporizes on inhalation. So air that leaves the lungs saturated with water at body temperature can be exhaled, finally, near outside temperature and thus with much less water vapor in it.

LIGHT BREATHING

Panting, as done by hot dogs and some other mammals and birds, puts this dead space to another use. Extra breathing will dispose of extra heat, just as perspiration does but without losing salts inherent in the perspiration process. If extra breathing moves the tidal volume of the lungs in and out, however, it would cause too much carbon dioxide to be removed from the blood. That reduces the acidity of the blood to intolerable levels. Prolonged deep breathing by a nonexercising human can cause loss of consciousness and other problems.

Furthermore, the additional work of extra breathing generates as much heat as one can lose by the activity. What dogs do fixes both problems. Panting depends on very shallow breathing, exchanging air almost entirely within the dead space. And it takes little extra work because it is tuned to a frequency matching the natural elastic recoil of an animal's thorax. That is why dogs run with their mouths open and their tongues hanging out.

EXTRA SACS

Birds evade another of the problems of reciprocating ventilation. They may breathe in and out, but this breathing inflates and deflates air sacs, rather than the lungs themselves. Valving within the system then pushes a flow of air in a single direction, although at a somewhat unsteady rate. As a result, bird lungs are small enough to carry into the air for flight, but efficient enough to supply the large and oxygen-hungry flight muscles.

▼ When breathing in and out, we cannot exchange all the air in our respiratory systems, so fresh air must inevitably be diluted by air from which oxygen has already been extracted.

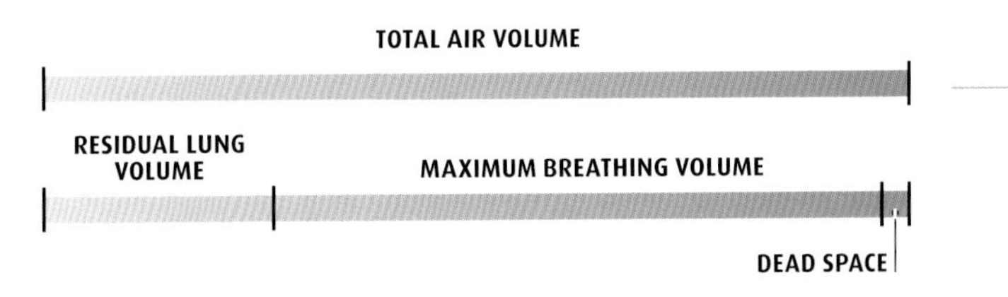

GOING ONE WAY

Human technology rarely bothers with reciprocating ventilation, relying instead on our fine unidirectional pumps and blowers. We do, however, occasionally use a purely in-and-out flow system, such as meat basters and hypodermic syringes, designing them as best we can to have minimal dead space.

Valve arrangements such as those in the respiratory system of birds find wide application when we pump fluids. The most common example is a bicycle pump, where a pair of valves converts the in-and-out movement of air beneath its piston into a pulsating but unidirectional flow into the tire.

▶ Panting dogs avoid expelling too much carbon dioxide, which would make their blood too alkaline, by very shallow breathing, exchanging only dead-space air.

Pumps for Ventilation

When winds and water currents will not do the job, organisms use pumps to force fluids through themselves or their dwellings. Pumps differ widely, chosen by nature's usual combination of functional demands and ancestral legacy. Most impose some energetic cost. All impose some cost of construction.

Engineers often divide pumps into two types: positive displacement and fluid dynamic. Most positive-displacement pumps persuade fluid to fill some kind of chamber, then persuade that chamber to become smaller, which makes the fluid leave through the intended exit. They invest their power more to boost pressure in the fluid and less to propel volume. The bicycle pump is an example.

Most fluid-dynamic pumps, by contrast, put kinetic energy directly into the fluid with an impeller. They produce relatively lower pressures, but push out greater volumes. A fan in a pipe is perhaps the simplest form of this type of pump. A blow-dryer may be the most familiar.

NATURAL PUMPS

The same distinction between pumps works for natural pumps as well. And the parallels tell us a lot about how nature, like engineers, matches pumps with their tasks.

If pressure change is the criterion, then the evaporative pump with which trees draw water up from roots to leaves tops the list of nature's positive-displacement pumps. Pressures inside the tree may in extreme cases exceed 100 atmospheres. Even the more typical 10 atmospheres represents a far higher pressure—if a vastly lower volume flow—than exists in any vertebrate heart, which tops out at 0.5 atmospheres. Nature's osmotic pumps also generate remarkable pressures, commonly around 10 atmospheres.

Muscular tube- and chamber-squeezers produce still lower pressures. They come in several kinds—some (such as our hearts) have valves just like those of bicycle pumps, while some use traveling peristaltic squeezes (our intestines), and a few depend on some kind of dynamic valving (bird lungs).

◀ **Trees use powerful positive-displacement pumps capable of sucking water up through tiny pipes high above ground.**

▲ Bees line up in front of a hot hive, beating their wings as fluid-dynamic pumps, and fan-cool air into the hive. A human-technology equivalent is the blow-dryer.

◀ A bicycle pump is the most familiar kind of positive-displacement pump. It contains a chamber whose volume can be varied with two one-way valves.

PUMP INNOVATIONS

Human technology has not yet harnessed the wonderful evaporative pump found in every tree, which can suck water from almost dry soil, then hoist it through tiny pipes tens of yards above ground—with no moving parts.

Our peristaltic squeeze pumps are inefficient and rare, but related moving-chamber pumps range from the dragon-bone chains that irrigate some rice paddies in Asia to the gear pumps that circulate oil in contemporary automobiles. Valve-and-chamber analogs of vertebrate hearts are common as both hand-operated and motor-driven devices.

FLUID-DYNAMIC PUMPS

Nature's fluid-dynamic pumps include very low pressure ones such as the flow inducers mentioned earlier, with pressures of a hundredth of an atmosphere. Ciliary pumps produce pressures on average only slightly higher. Since cilia can get longer but not thicker, ciliary pumps do not scale up well, so muscle runs almost all of nature's large dynamic pumps, with paddles and propellers providing pushes. Examples include paddle ventilators of the burrows of some worms and crustaceans, and multistage blowers formed by lines of honeybees beating their wings to ventilate hives.

Natural and human technology differ most in fluid-dynamic pumps. Our toolbox lacks anything close to cilia or muscles, the predominant engines of animals, but nature lacks our versatile and efficient wheel-and-axle devices. So, while the general category is found in both technologies, the specific devices are remarkably divergent.

COMPARING APPROACHES IN NATURE AND SCIENCE

Comparing pump technologies raises two issues. First, why are some of the most effective pumps rare in nature? Among simple positive-displacement pumps we find no air-lift pumps such as the ones we use to aerate aquaria. Maybe underwater organisms with gas bladders find their gas too precious to expend in pumping. Among fluid-dynamic pumps we find no centrifugal ones like those of washing machines, dishwashers, and automobile cooling systems. That is easier—nature makes no wheel-and-axle devices.

Second, why do some kinds of pumps in nature have no use in human technology? We have never used suction, whether generated by evaporation or any other means, to lift water above the 10-yard (10-m) limit set by atmospheric pressure. Cavitation in anything but microscopic tubes seems an insuperable obstacle. For our valve-and-chamber pumps we rarely capitalize on flexible materials—lacking anything like muscle, we mostly use moving pistons in cylinders, rather than squeezing chambers. With no cilia, we make no pipes that use their walls to pump. So nature still holds general lessons, illustrative examples, and a catalog of opportunities for us.

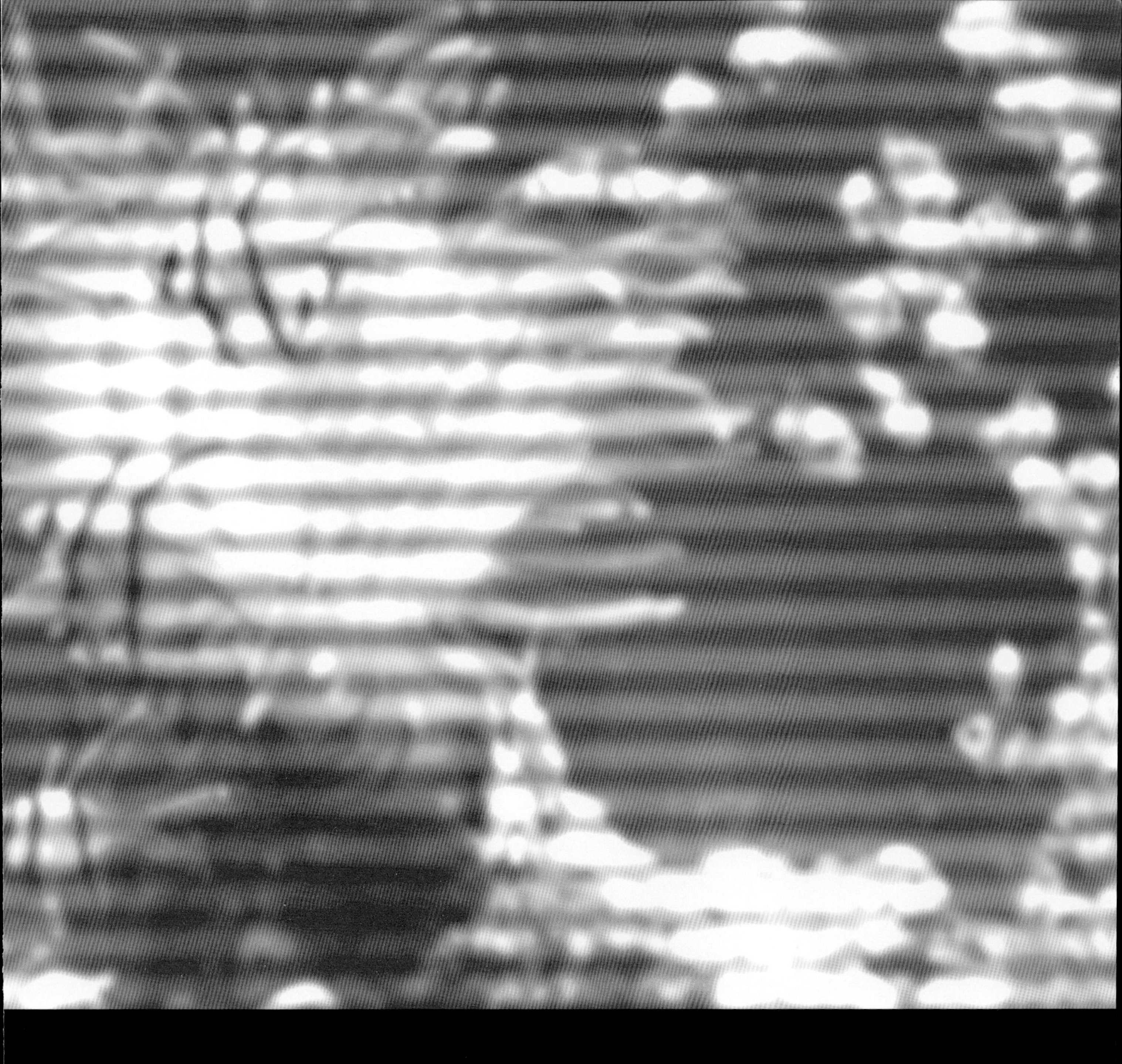

6 | NEW MATERIALS AND NATURAL DESIGN

Introduction

Biology, like technology, is reliant on materials for the structures it makes. These structures also have to be cheap and reliable. Evolutionary fitness (and therefore survival) is, in part, value-based—the survival of the cheapest. Success requires the ability to compete for, and survive upon, resources that may be scarce.

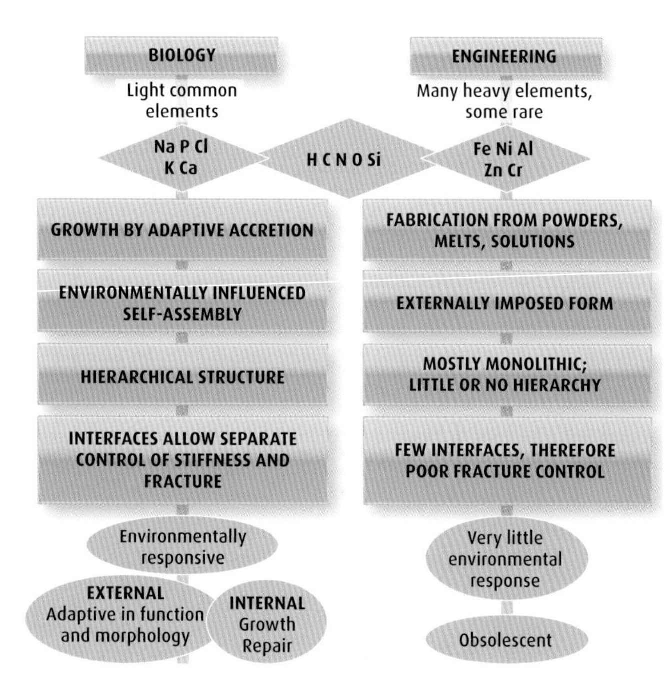

Animals and plants are largely constructed from materials that are readily to hand, such as water, carbon, hydrogen, nitrogen, calcium, and silicon. Few metals are used, although iron, zinc, and manganese are essential in various ways. Scientists at Cambridge University, UK, led by Mike Ashby, have produced some "property maps" of materials, showing the relationships between various properties of materials in the design of structures. The map that plots stiffness and strength against one another (opposite) is of use when designing load-bearing beams, sheets, and struts. When density is taken into account, natural and human-made materials cover much the same area, indicating that they deliver the same mechanical properties. Biological materials, however, are constructed at ambient temperatures and use only two polymers (protein and polysaccharide), with two ceramics (calcium salts and silica), and a few metals, whereas human-made materials require high temperatures and hundreds of polymers.

Most significantly, biological materials contain water, a component that is missing from nearly all human-made materials and which is viewed in that context as an agent of degradation. Yet water is the cheapest material available. It provides the medium in which all the chemical reactions of biology take place, and is crucial in controlling the "self-assembly" of the resulting materials. Water acts as a plasticizer, giving polysaccharides and proteins space in which to move and inhibiting (or masking) many of their interactions, thus

◀ Engineering (right column) and biology (left column) are very different in their raw materials and the way in which they use them.

▲ Beetles are the most numerous of all insects. The hard shell—cuticle—is a sophisticated composite, somewhat like fiberglass.

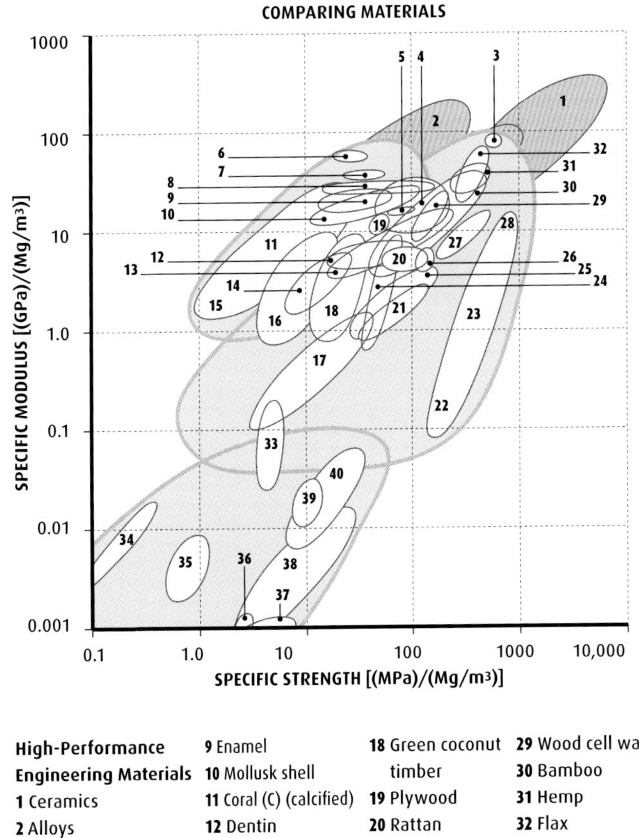

▶ When density is accounted for, there is little difference between the modulus (i.e. stiffness) and the strength of biological and engineering materials. The biological materials (light blue; specific materials in yellow) cover and obscure nearly all the engineering materials (dark blue).

High-Performance Engineering Materials
1 Ceramics
2 Alloys
Biological Materials
3 Chitin, cellulose
4 Wood (II) (parallel to the grain)
5 Dry coconut timber
6 Calcite
7 Aragonite
8 Hydroxyapatite
9 Enamel
10 Mollusk shell
11 Coral (C) (calcified)
12 Dentin
13 Compact bone
14 Keratin
15 Coral (T) (tanned protein)
16 Wood (T) (across the grain)
17 Cancellous bone
18 Green coconut timber
19 Plywood
20 Rattan
21 Wool
22 Viscid silk
23 Cocoon silk
24 Cuticle
25 Collagen
26 Antler bone
27 Cotton
28 Dragline silk
29 Wood cell wall
30 Bamboo
31 Hemp
32 Flax
33 Cork
34 Muscle
35 Parenchyma
36 Reslin
37 Cartilage
38 Elastin
39 Skin
40 Leather

increasing their range of mechanical properties. It is also a valuable structural component for resisting compression—for example, in plant cells, where it is contained at pressures around ten times atmospheric pressure; in materials such as cartilage, where it is "bound" chemically into the structure; and in types of worm, where it performs a similar function to bone. It would be advantageous if we could develop a system of materials synthesis and processing based on water because water would then cease to be the disruptive agent that it currently is. This would rule out processes requiring high temperatures, such as metal and ceramic production, but nature manages very well with a wide range of polymeric materials such as silk and wood that are extremely durable when cared for properly.

IF IT'S SCIENCE, THERE MUST BE NUMBERS

If there are no numbers, there can be no comparisons and therefore no science and no improvements. So we need to know about such things as stiffness, which is the resistance of a material to deformation (not be confused with strength, which is the resistance of a material to breaking), and which is measured in newtons per square meter. A "newton" is (roughly) the force exerted by a large apple in the Earth's gravitational field. A single newton acting over an area of one square meter is known as a "pascal." A table gelatin dessert has a stiffness of about 1,000 pascals (1 kPa); the stiffness of rubber is about 1,000 times this (1 MPa); and a stiff plastic such as polymethyl-methacrylate (known as Perspex or Lucite) is about 1,000 times (1 GPa) as stiff again. Different types of insect cuticle can cover this entire range.

Water-based processing in biological materials can be productive. For example, the cuticle or exoskeleton of insects is composed largely of fibers of chitin (a polysaccharide polymer) in a protein matrix. Silklike structures in the protein interact with similarly spaced side-chains on the chitin. When the protein is laid down, it contains a significant amount of water; when the cuticle stiffens into a load-bearing skeleton, this water is removed, allowing formation of bonds between the protein chains. Despite its watery origins, cuticle can be very waterproof and resists degradation in wet soil for thousands of years. Plastics, the technological equivalent of cuticle, are not processed in this way, and need special treatment to produce materials with the high performance of the strength-to-weight ratio of insect cuticle.

Proteins

Proteins are made up of chains of amino acids. Although there are hundreds of amino acids, only about 20 form natural proteins. Their wide range of chemical properties, however, provide many ways for proteins to interact with each other and with their immediate environment.

▶ Collagen, a protein fiber, is the body builder of soft skeletons.

The shapes of proteins are controlled by the amount of movement allowed by their components. The smallest of the 20 amino acids, glycine, allows most movement and so gives the protein chain the most flexibility. Proline, on the other hand, has a structure that provides rigidity and consequently stabilizes the chain. Amino acids can be hydrophilic (able to form bonds with water) or hydrophobic (repelling water).

◀ The threads of the mussel's "beard" were used to make very strong purses, hence the scientific name *byssus* (derived from the name of ancient Egyptian fine linen, or byssus cloth).

SOME BASIC SHAPES AND PROPERTIES

Collagen is composed of three protein chains that twist around each other. The chemistry and size of the amino acids control the shape of the chain. The tensile stiffness of collagen fibers, about 1.5 GPa, is typical of amorphous plastics, but is low for a relatively ordered material such as collagen, in which the forces are borne by the main chain. A stiff and strong form of collagen is found, however, in the wall of the hydraulic capsule that powers the sting of sea anemones and jellyfish. The stiffness of the collagen here is about 25 GPa, allowing the contents of the capsule to reach a pressure of 150 atmospheres before the dart shoots out, such that some jellyfish stings are able even to penetrate the shells of crustacea.

Keratins are typical of the outer covering of vertebrates. In mammals they form hair, quills and spines, horns, hooves, baleen, and the outer layer of skin. Mammalian keratin structure resembles coiled springs (with a stiffness of 6–10 GPa); the keratins of birds and reptiles are constructed from

▼ Spiders' webs are made of silk, a fiber stronger and stiffer than steel.

▶ The stings of sea anemones are powered by high-pressure liquid in an hydraulic capsule.

an extended twisted sheet. Keratin is always found inside the cells that produce it, which are closely packed parallel to each other and stick together. The mechanical properties of the springlike structures can be deduced by mechanically testing hair or horn to produce a graph that describes the relationship between stress and strain.

Silk is stiffer and tougher than most engineering materials, especially when density is taken into account. It is produced by many moth caterpillars to make cocoons, and by spiders for making webs, and it is largely made of protein with a sheetlike structure. Silk can have very high stiffness and strength because the covalent bonds (the stiffest and strongest bonds in chemistry) are orientated in the direction of the fiber. Across the sheets the bonds are weaker, so that the protein chains can slip past each other quite easily, giving a fiber which, although stiff, is very flexible.

Elastin is found in all bony animals. It is a soft yellow material that is stretched between the bones of the skeleton, forming ligaments such as the ligamentum nuchae at the back of the neck, which acts like the counterbalance spring in an anglepoise lamp to support the head. Elastin is also found in the aorta and arteries, where it provides elastic recoil to even out the pulses of the blood and reduce the load on the heart. Elastin can be prepared in the laboratory by heating a ligament to 230° F (110° C), at which temperature the collagen and other tissues melt and dissolve, leaving the elastin untouched (because of the heat-resistant cross-links that hold elastin together). The most significant work on elastin was done by Dan Urry at the University of Birmingham, Alabama. He showed that elastin fibers are stretchable helices with a watery core. Elastin has the strange property of contracting abruptly when chilled, and so it could be used as a molecular heat engine to power tiny machines.

The threads anchoring mussels to rock contain collagen, silk, and elastin, mixing mechanical properties. They are strong and stiff, but can stretch to accommodate waves, which pull the threads parallel so that other threads are brought into line. This increases the strength of the anchorage. The glue holding the thread to the rock has been studied to produce a glue for medicine (see also Chapter 1).

Polysaccharides—Structural Sugars

Polysaccharides are a type of carbohydrate, or sugar, molecule that form the basic structural materials for many living things. The sugar units link to produce stiff fibers or watery, space-filling gels. Examples are cellulose fibers in plants and the cartilage found at the joints between bones.

▶ Vines are fibrous and strong in tension, right down to the molecular fibers of cellulose from which they are made.

The molecules in these polysaccharide chains are packed together in crystalline nanofibers that are a few tens of nanometers across but can be as long as you like. The stiffness of a cellulose nanofiber is 135 GPa and that of chitin nanofiber at least 150 GPa, which is comparable with carbon fibers. Since cellulose is part of our chemical makeup, it can be used to create replacement tissues for medical applications that are clinically safe. The nanofibers are laid down in patterns reminiscent of liquid crystals, in which short and stiff molecules lie closely against each other like bundles of spaghetti, making larger and stiffer structures.

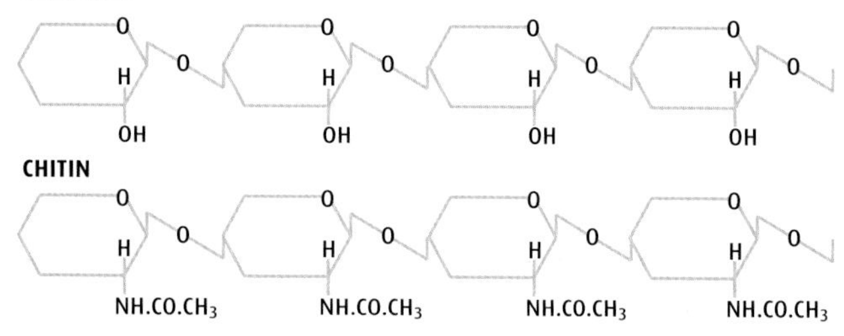

◀ Cellulose (forms part of our chemical makeup) and chitin (forms insect cuticle) are both made from strings of sugar. They are very similar in structure, very common, and very strong.

In fact, all self-assembly (and thus growth) can be interpreted as a form of liquid crystallinity in which the proportions of polysaccharides and water can vary to yield a wide range of mechanical properties, depending upon the orientation of the fiber and the amount of water in the protein matrix. Insect cuticle, for example, can be as soft as sputum (stiffness of about 1 kPa) or stiffer than fiberglass (about 20 GPa).

THE STRUCTURE OF CELLULOSE

The best model for the synthesis of cellulose was suggested by R. D. Preston, a professor of biophysics and botany at Leeds University, UK, who proposed that cellulose is spun from the center of rosette-shaped enzymes floating in the cell membrane. It was an idea for which he had no real evidence, but it stimulated a search for a structure. The enzyme rosettes were subsequently found, arrayed hexagonally in groups of 100 or more, wandering around the membrane and leaving behind them a trail of cellulose nanofibrils that aggregate into microfibrils.

The orientation of cellulose in the cell walls is guided by a network of microtubules arranged on the inner surface (cortex) of the cell. The shape, size, and stiffness of the growing shoot of a young plant is governed by the mechanical properties of the outer wall of the outer layer of cells. This wall is several times thicker than any of the other cell walls in the growing shoot, and so has a greater mechanical influence. The orientation of the microtubules can be changed by external stimuli such as light, auxin (a plant hormone), and mechanical strains such as those due to bending. These stimuli are additive, so, for example, a small amount of auxin makes the cells more sensitive to the other stimuli. At the same time, the growth rate changes, and so it is the reorientation of cellulose microfibrils, with the changing orientation of the microtubules acting as a go-between, which governs growth and form.

The cellulosic network is not the only one. Pectin (another polysaccharide, known for its gelling, thickening, and stabilizing properties in food preparations such as jams) is present in most primary cell walls and is particularly abundant in the non-woody parts of terrestrial plants. It is also present in the middle lamella between plant cells, where it helps to bind cells together. Pectins have many and large side-chains that allow them to fill spaces between microfibrils and cells, and stick cells together. In ripening fruit, the pectins are made soluble and cease to stick so well, causing the texture of the tissue to change. Another network is made of lignin, a complex and inert polymer of alcohols based on hydrophobic, highly stable hydrocarbon units (known as "phenol rings") that both limits the mobility of the cell wall fibers and makes the walls drier and stiffer.

SPACE FILLERS AND GELS

Polysaccharides can also make space-filling structures. Unlike proteins, which have only one type of linkage between the subunits, the 6-carbon sugars of most polysaccharides can be linked through any and (nearly) all of the carbon atoms. They can therefore branch and produce florid shapes. They also bind large amounts of water, to the extent that a stable gel can be formed with less than 2 percent of solid material. In animals these gels are combined with collagen and other proteins to make cartilage, the translucent white material that makes up joint surfaces and the "bones" of sharks. Similar materials, containing even more water, make the skeletons and shapes of most small marine organisms.

Most of these ways of designing materials and structures are not applicable to human-made materials, largely because we do not understand how to use water as a structural unit. Our materials technology is mostly derived from chemicals found in oil, and oil and water do not mix because they are so different. The closest human-made analogies to biological materials are fibrous composites such as fiberglass, but biological materials are far more complex and difficult to understand. If we could work out how to replicate these organic processes, it would potentially be very cheap to make similar structures such as structural framework and enclosures, and waterproof casings and shells, because they would use a very small amount of solid material.

▶ A ripe plum is soft because the "glue" between the cells has partially dissolved, and the cells can slide past each other.

Strange Liquids

Water is the most important material in biology, yet the strangest and least understood. It provides the medium in which the molecules of life assemble themselves and in which they interact. It softens proteins and polysaccharides, yet provides the skeletal support for most plants and animals.

Many gels (including food gels such as ice cream and gelatin dessert) are remarkably dilute—only about 1 or 2 percent solids—and yet are stiff and stable. How can such a small amount of material command the attentions of so much water? Current wisdom is that water is a small molecule and so can exert its influence over only small distances, especially when there are salts present that can affect these interactions. If this is the case, how can stable but dilute gels come about, and how can a coherent raindrop form around a small particle? Gerry Pollack, a biophysicist at the University of Washington in Seattle, discovered that hydrophilic surfaces affect the organization of water molecules for distances of hundreds of micrometers (μm). Such surfaces can create a zone from which "large" particles such as spheres of diameter 2 μm

▼ **Water is a basic essential for life, yet is still mysterious in its ways. Nature uses it as a medium in the structure of many kinds of materials.**

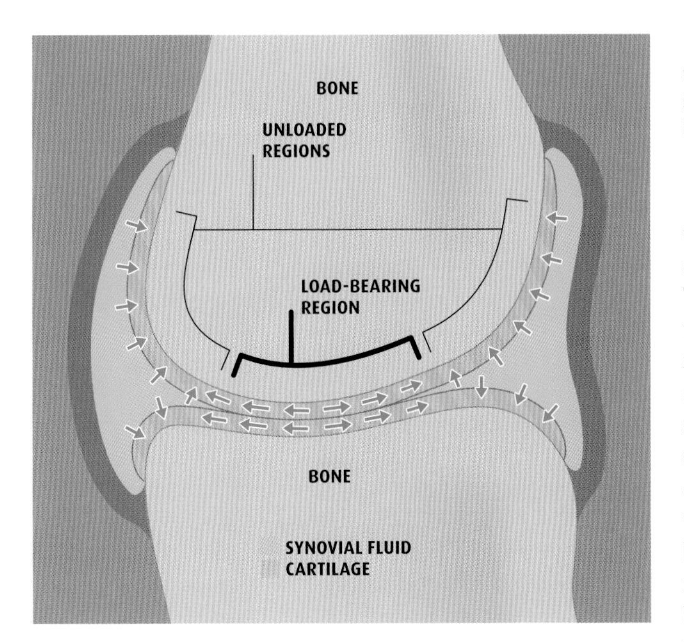

◀ The synovial membrane around the knee joint retains the lubricating synovial fluid, which is absorbed into the soft cartilage in areas where there is no contact between the articular surface, but which is squeezed out where there is a loaded contact.

(this is at least 100 times larger than a "normal" molecule in solution) can be excluded. This exclusion zone can easily be seen with a light microscope; it develops in a few minutes from a stirred solution/suspension of particles. The water in this organized exclusion zone can flow, although slowly. Whether or not the exclusion zone could be maintained against external forces has not been examined experimentally, but it would be interesting from the point of view of lubrication.

ADVANCED LUBRICANTS

Current theories of biological lubrication suggest that large molecules (usually of hyaluronic acid) are tethered to a surface. This is what is understood to happen in synovial joints such as the knee and the hip. As the joint is loaded, a watery solution is forced out of the cartilage. The hyaluronic acid, sitting like the hairs of a brush on the cartilage surface, binds large amounts of water as a gel and provides the lubricating separation between the two surfaces. The very low sliding friction at natural synovial joints has not yet been attained in any artificial watery joints.

A similar effect has been found using a thin film, less than a nanometer thick, of a large polysaccharide from the alga *Porphyridium* sp., adsorbed from aqueous solution. The friction with this experimental system was low and there was no wear even at high pressure. Atomic force microscopy showed that the biopolymer was fixed to the rubbing surface, but was mobile and easily dragged upon shearing. The adsorption of this polysaccharide onto surfaces, its low friction, its robustness, and the weak dependence of the viscous friction force on the sliding velocity make it an excellent candidate for use in water-based lubricants. (In the world of engineering, viscous friction forces are proportional to velocity difference between the sliding surfaces; therefore, as the velocity increases, so does the friction force.) It could even be added to synovial fluid in joints such as the knee. That such a thin film can have such a dramatic effect suggests that the water is being organized by the polysaccharide, giving a basis for lubrication very different from the classical explanation, which is based on thick "polymer brush" layers. So how much of the lubrication is down to controlled shearing of the liquid crystalline structure of the water? How is its shear strength affected by the molecules stabilizing it? These observations—on the nature of water at surfaces and the way that biological molecules can influence it—are of great importance. We are entering a new era of understanding of this ubiquitous material.

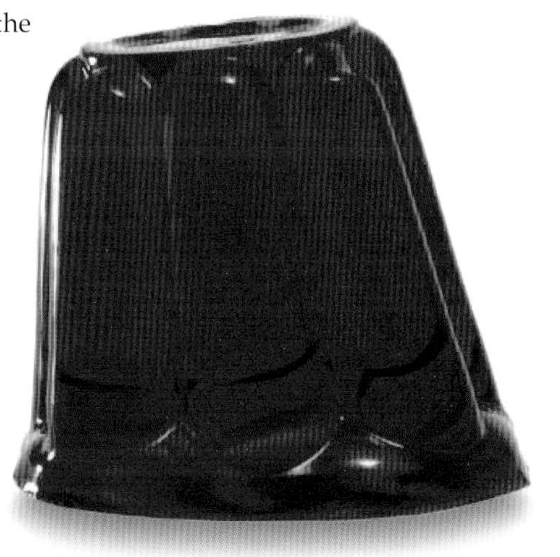

▶ Gelatin dessert consists of up to 98 percent water and very little solid material—so why does it not flow away?

Stiffness Out Of Water

Water cannot easily be compressed and so can transmit a force, as a bone can, but only if it is contained in a stiff-walled balloon. In worms and maggots, the entire animal is a low-pressure balloon. In plants, the balloons are cells 0.004 inch (0.1 mm) in diameter with internal (turgor) pressure of 10 atmospheres or more.

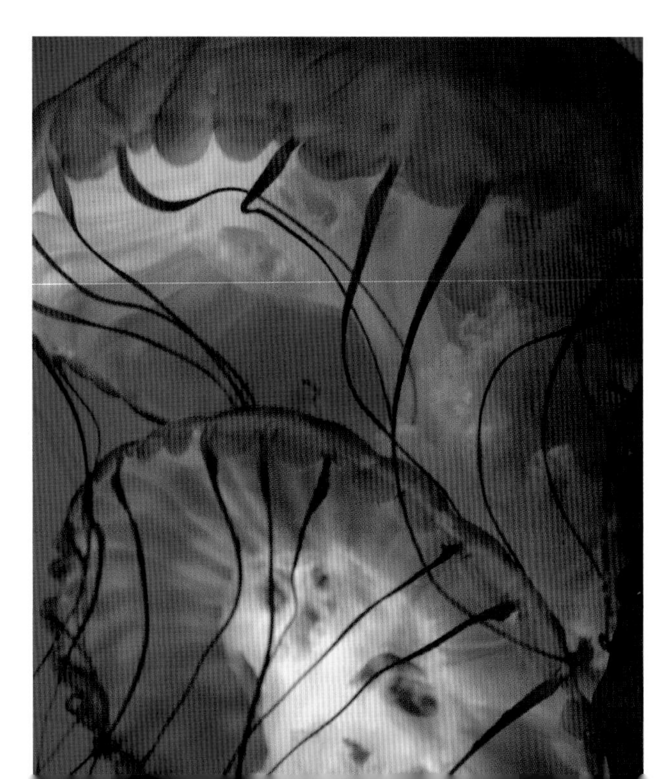

In plants the cells stick together and transmit forces one to another, giving a texture characterized as "firm" in fruit such as apples, and "stiff" in non-woody plants such as the stem of the dandelion. This flower stem has to withstand the force of the wind, and its own weight is trivial. The critical factor in maintaining its upright stance is its ability to withstand the compressive forces experienced on the concave (downwind) face when it bends; the stem will never fail on the side bending away from the wind because the cellulose is so strong in tension. The most likely mode of failure is local buckling of the fibers in the cell walls, leading to a compression crease.

TURGIDITY

A flower stem has small thick-walled cells on the outside and large thin-walled ones inside. As the turgor pressure increases, the inner cells are liable to stretch more than the outer ones, and a strip along the stem curls, with the thinner-walled cells on the outside of the curl. This is because the turgor is providing a constant force, so the stress will be greater on the thinner walls. An intact stem will not curl because all the "strips" are pulling against each other, resulting in circular or "hoop" stresses in the outer layers, thus any side load will have to pay off this high stress on the compression side. If the plant stem can concentrate its

◀ Jellyfish are basically sea anemones turned upside down. The jelly is a soft skeleton, against which the muscles can react.

▲ The dandelion flowering stem shown above was split along its length and put in tap water. The cells absorbed water and expanded.

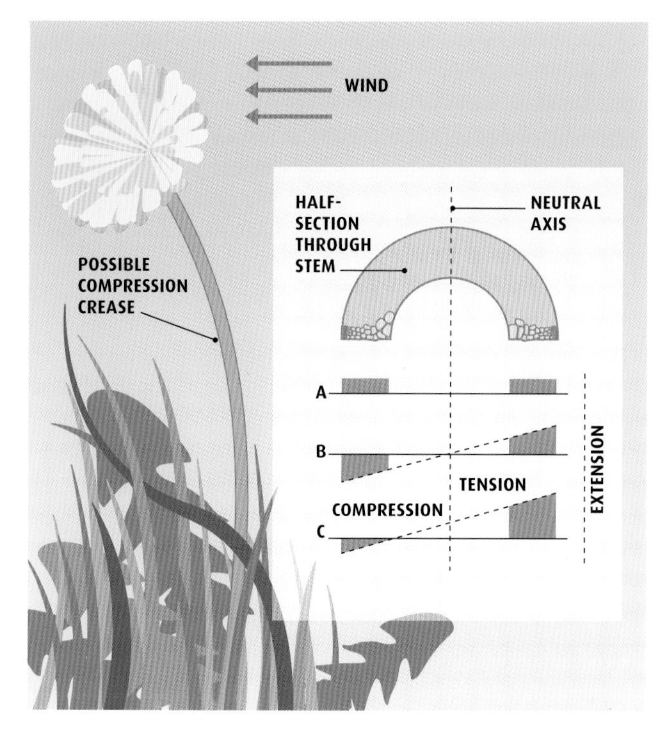

◀ Turgor pressure stretches the cells in the stem (A), and wind load stretches one side of the stem and compresses the other (B). Add them to get (C).

cell wall material at the periphery of the stem, the tensile force in the material will be at a maximum just where the compressive stress on the structure will be highest. Admittedly this mechanism puts much greater tensile loads on the tension side of the stem, but the cell-wall cellulose can take stresses of the order of 1 GPa before it breaks. An advantage of this design is that the small thick-walled cells that take the highest compressive stress are much less likely to fail by buckling than the larger thinner-walled cells found nearer the center of the stem. Thus the gradient in cell-wall thickness across the stem has a twofold effect in resisting buckling, acting both at the level of the cell and at the level of the entire structure. We use a similar approach in making struts and beams that have to withstand bending or compression, distributing most of the material around the outer edges of the structure. But once again the detailed design that nature uses eludes our relatively crude assembly processes, and we have to use more material in order to achieve the same mechanical result.

THE TUBE AS A MECHANISM

In animals the pressures are much lower because they are much more active than plants and have to be able to change their shape rapidly and easily. A tube is a very common shape for animals and bits of animals (worms, guts, blood vessels, tissue surrounding muscles and nerves), and is constructed in a very similar way throughout nature. Biological tubes need to be able not only to bend, but also to change length and diameter. The universal solution to this problem is one that is familiar to the designer of pressure vessels—crossed-helical fibers of collagen, like the structure of fishnet stockings. In the latter, it is readily seen to adapt itself to a variety of changing shapes. A tube of constant diameter has the angle between the fibers at 54° 45′ when at maximum volume. If the tube is held below the maximum volume, its shape can vary from short and fat to long and thin; however, material properties come into play in, for example, the tubular propulsion mechanism of the chameleon's tongue. This elongates the double-helical tube at constant diameter, thus stretching the collagen fibers and storing elastic energy in them, which then catapults the heavy tip of the tongue out of the mouth. The properties of the helically wound tube have been known for many years and used in the design of pressure vessels such as cannon and rocket engines. The idea of using the principle as a mechanism capable of controlling movement seems largely to have been ignored, and it is commonly found only in the sheathing of wires in electrical devices.

▶ The chameleon's tongue is catapulted by stretched collagen, thereby extending its reach when capturing prey.

Soft Composites

Skin defines the shape of an organism, separates and protects it from the outside world, yet allows all forms of communication (senses, nutrition, respiration, excretion) to continue unimpeded. Mammalian skin is mechanically one of the most complex, and insect "skin" (the cuticle) is one of the most versatile.

The dermal layer of human skin is a network of collagen fibers with some elastin, in a protein–polysaccharide matrix. The outer surface (epidermis) is protected by a layer of cells (the stratum corneum) full of keratinous protein in springlike helices. The mechanical properties of human skin are in general nonlinear, anisotropic, and dependent on strain rate. The orientation of the collagen fibers was explored around 1880 by the Austrian anatomist Karl Langer (1819–87), who punched circular holes into the skin of cadavers and noted that the holes became oval. He drew lines connecting the long axes of the ovals, thus mapping the directions of lowest stiffness. As the skin is stretched in any direction, it gives rise to a J-shaped stress–strain curve that is typical of nearly all collagenous structures and is due to reorientation of the collagen fibers toward the direction of stretching. The orientation of collagen in skin varies in all three directions, although the tendency is to model the properties in only two dimensions—in the plane of the skin. Since the reorientation of fibers in skin seems to need relatively little force (although quite large displacements), it is difficult to decide what should be the starting point for any mechanical test.

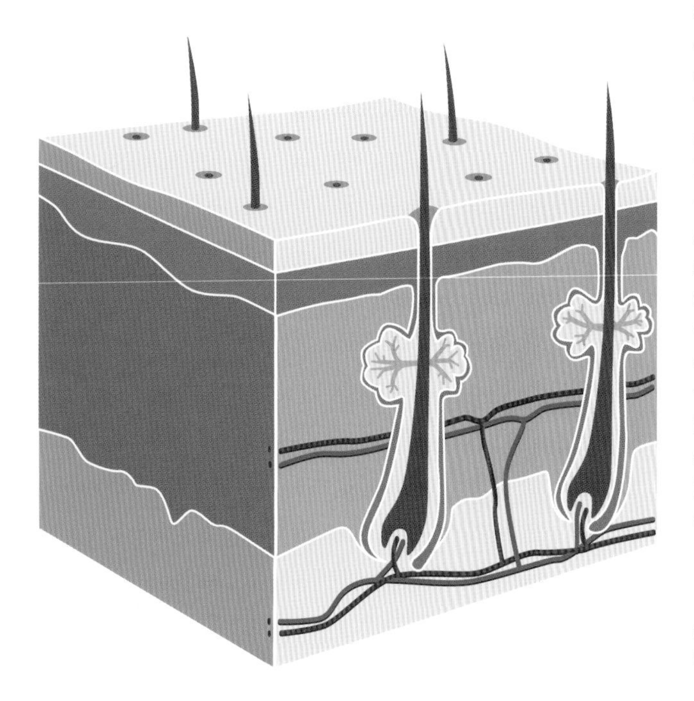

◀ **Skin is very complex mechanically and is made of several layers. As the outer layer is shed, more cells from underneath replace it.**

TOUGHNESS IS ONLY SKIN DEEP

In life the skin is rarely stretched in only one direction, but rather in a variety of directions all at once. In experimental terms this means stretching it in ten or more directions at once with a ring of attachments. Even the most sophisticated computer analysis cannot produce a general model of the mechanical behavior of the skin. For example, in a finite element model, the material or structure under examination is divided up into a large number of defined (finite) elements whose properties can be expressed in simple terms, as can their interaction with neighboring elements. This allows complex shapes with nonlinear properties to be modeled as an assemblage of simple shapes with linear properties. Skin, however, defeats the engineer (who is used to repeatable samples for testing), in that there are differences in the relative stiffness and extensibility of skin from different individuals that cannot be related to gender, location on the body, or environment. Under these circumstances, the experimentalist is commonly driven to use a "constitutive equation," an equation derived more from observation than from theory, although theory can follow. When Robert Hooke expressed his law *ut tensio sic vis*—as the extension, so the force—he did not know where the relationship came from, and so this important and useful formulation is constitutive. The problem therefore is that one does not know whether the description includes all the important factors. It deals with correlation, rather than causality.

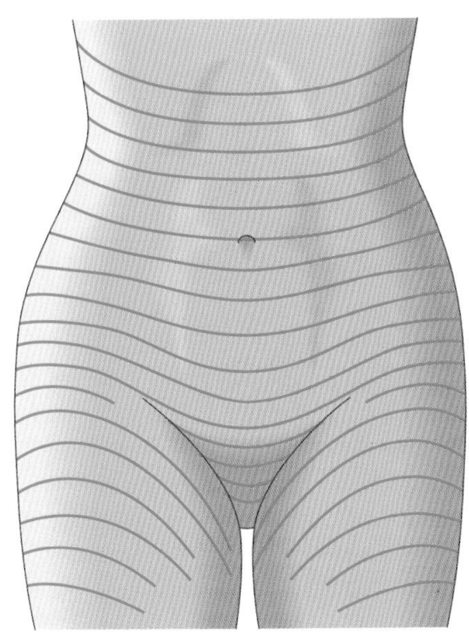

▲ The collagen fibers in skin are oriented so that the body can move freely inside it. These are referred to as "Langer's lines."

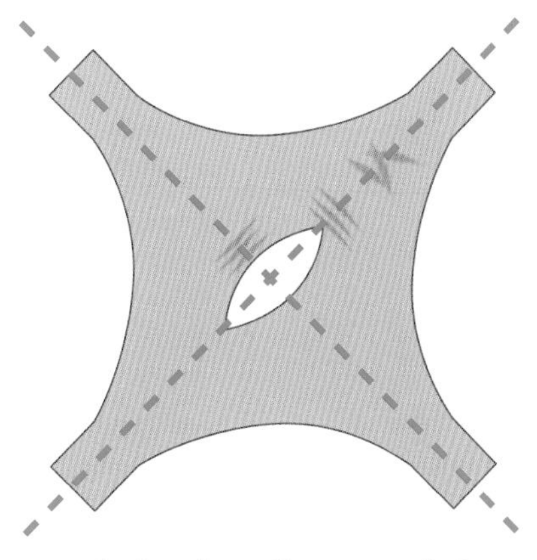

▲ The collagen fibers can reorient in response to a wound. Here, they are stopping the slit from extending farther.

Several factors combine to make skin a tough material. Low stiffness at small extensions means that deformation of the skin can be localized, isolating damage. If the skin is damaged, as with a knife, then the release of tension occasioned by the presence of a wound causes the fibers at the ends of the cut to be reoriented across the direction in which the wound might be expected to develop, giving the skin an adaptive response to damage. The closest we currently come to this sort of material are knitted fabrics that are very difficult to tear, or plastics with long fibers that are used to make toughened envelopes. But these were designed independently of nature's prototypes, and the economics of industry are such that, even if a better idea comes along, the decision as to whether it is adopted is more due to the expense of retooling than to possible improvements in the product.

Skin is not the only soft fibrous material. The various tubes that carry air and liquids around the bodies of animals, ranging from worms to humans, are very similar to inner layers of skin in their components, although the orientation of the fibers is closely related to the mechanical requirements. Internal connective membranes, which support and confine the organs, are more delicate. The main component of all these tissues—collagen—makes up a quarter of the total protein found in the body.

Stiff Composites

Water is essential for the manufacture and assembly of biological composites. One way to make a stiff material is simply to get rid of the water and allow the stiff fibers to stick together. In plants and animals, this is achieved by chemistry in a process that is very similar to tanning leather.

▲ This female locust (*Locusta migratoria*) is stiff like most insects. It and many insects are brown or black, both a result of a process known as phenolic "tanning."

Insect cuticle is a relatively simple composite material. The variables are the amount of chitin, its orientation, the type of protein, its hydration, and the amount and location of other substances (melanins, zinc, manganese, some salts). Stiffness ranges from tens of GPa in "hard" cuticle in the main body structures of the larger beetles, to 1 kPa in the softest cuticle of the extensible intersegmental membrane of the female locust. Cuticle is typical of hydrogen-bonded materials in that a very small change in water content of only a few percent can lead to an extremely large change in stiffness. The cuticle goes through a transition in stiffness at a water content of around 25 percent, suggesting that free water (which is sufficiently mobile to plasticize the protein) is being lost, leaving only a single layer of more tightly bound water on the surface of the protein that is insufficient to lubricate the movement of protein molecules relative to each other. This sort of transition is found in other systems, notably starch-based foods. It is also the basis of some of the more remarkable tricks that cuticle can play.

One of the most obvious of these tricks is the stiffening of cuticle after an insect sheds its old exoskeleton. What causes this dehydration? In nearly all instances the answer seems to be a very general mechanism—the addition of phenolic materials.

PHENOLICS

When you drink strong black tea with no milk, or you drink "dry" red wine, your tongue feels rough and sticky. Phenolic compounds in the drinks are reacting with your saliva and driving out the water. This reduces the lubricating properties of the saliva and gives the roughness. Phenols color brown, especially when exposed to air, so many teas are brown, as are wood and many insect cuticles. Phenols are ring-shaped molecules of carbon (C) and hydrogen (H) with oxygen (O) added, which biology makes even more reactive with an enzyme. The resulting molecule is promiscuous and uncontrollable, linking with almost anything that is the least bit receptive to its advances. Once its chemical lusts have been satisfied, the water-binding sites in the material have mostly been masked, which, together with the essential water-repellence of the central ring of the phenol, dries everything out. As the water goes, the remaining water-binding sites are brought together and interact, rendering the material not only better cross-linked, but also insoluble because the water cannot penetrate the network that is held together by these secondary bonds acting in cooperation with each other. This chemically driven dehydration can happen underwater, which explains how the mussel's byssus thread, secreted as a watery collagenous fiber, can become strong and stiff.

UNDERSTANDING WOOD

Wood is brown so it is a safe bet that it has got lots of phenols in it. Lignin is a complex phenolic material that keeps water at bay, sticks the cellulose fibers together, and fills in any gaps. Lignin is like a plastic and softens at about 212° F (100° C), so heated wood can be molded and bent, keeping its new shape when cooled. But wood is more complex than an homogeneous plastic because it contains nanofibers of cellulose, which does not melt at such a low temperature. Wood is also more complex than its human-made relative, fiberglass, so materials science cannot predict all of its properties. The problem with many engineering models is that they do not include enough detail to cope with the regular and precise structures that biological macromolecules can produce.

The best way of getting around this is to build physical models and test them. Wood is simple enough for this to work. The cellulose is wound around the wood cell at an angle of about 15 degrees. You can model this with a helically wound paper art straw. When the cells are pulled, the tear goes helically along the length of the straw and the straw becomes thinner. When this happens in an assemblage of straws (which models softwood quite well), they pull away from each other, giving a much increased area of fracture, and hence increasing the likelihood that the fracture tear will be absorbed. This mechanism has been built into a novel glassfiber/resin composite that is, weight for weight, five times tougher than anything under impact and could be used in lightweight armor, where it would be capable of resisting high-speed impacts or knife attack.

UNDERSTANDING PHENOLS

Phenols are based on the benzene ring (1), which is usually depicted as a hexagon with a ring inside it (2). Benzene will not mix with water. If one or more of the hydrogens has an oxygen added to it, it becomes a phenol (3), which is soluble in water. If the OH groups are on one side as shown, the molecule is unbalanced (arrows) and chemically reactive. Removing the two Hs gives the quinone form (4)—even more unbalanced. This is now highly reactive. Adding more bits labeled R (made up of C and H atoms) makes the molecules more unbalanced (5, 6). When the =O atoms bind to other molecules, solubility goes down again, and the benzene rings drive any water out. So you get chemical dehydration that works underwater!

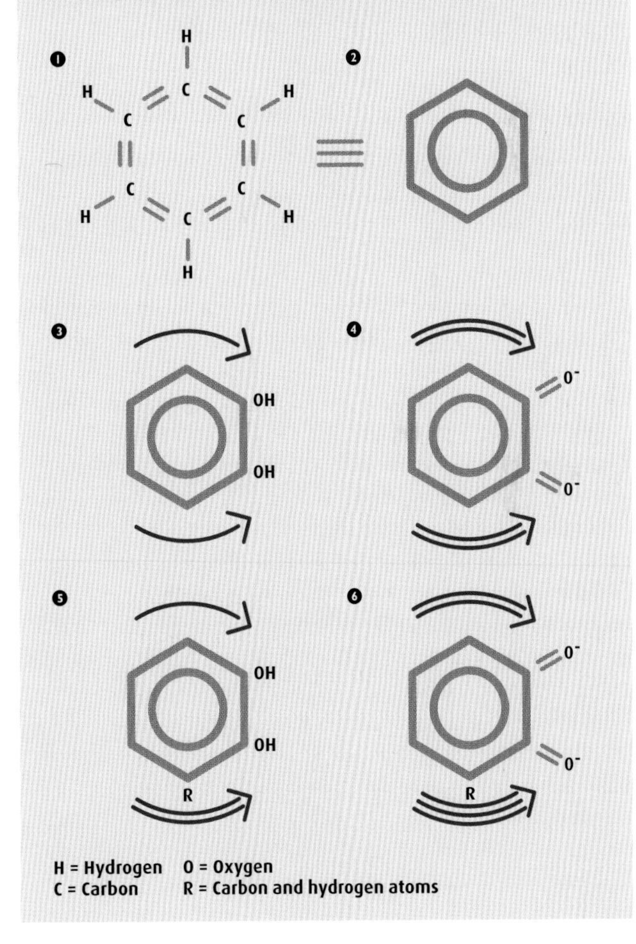

Ceramics in Snails

Some parts of animals or plants need to be strong and stiff for protection and support. But something like wood or insect cuticle, made of cross-linked proteins and fibers, costs energy. The components have to be gathered in the food, digested, and synthesized. Chalk and silica are stiffer and need less energy to make.

▲ The pearly nautilus shell mother-of-pearl is 3,000 times tougher than the chalk from which it is made. Even when it is breaking, the layers of shell are held together by silklike threads less than a micrometer (a millionth of a yard/meter) long).

The most abundant and easily used material is crystalline calcium carbonate, although phosphates of calcium, often magnesium, and in some instances opal, a hydratable form of silica, also occur. The crystals can be precipitated from a sufficiently concentrated solution and redissolved when the time comes. The difficulties come at the beginning and the end of the process—how do you get the mineral to crystallize where it is needed, and how do you stop the crystals from growing when they are large enough? The wonder lies in the degree to which such relatively simple inorganic materials can be fashioned and processed so that they become durable, tough, and incredibly precisely shaped.

TOUGHNESS FROM CHALK

Snail shells are an extensively studied calcification system. Biomineralization studies focus on mother-of-pearl, or nacre, probably because its geometry of layers of relatively large platelets—polygons about 0.3 inch (8 mm across) and 0.02 inch (0.5 mm) thick—is so beguilingly simple. The platelets are made of aragonite, a dense crystalline form of calcium carbonate. The platelets are connected together via small holes in the protein membrane that separates the layers. This neatly solves the problem of how to start each individual platelet growing as a crystal and how to ensure crystallographic, and probably hence mechanical, perfection from one layer to the next (it is just one big crystal). The matrix (glue) between the sheets of nacre is very tenuous (less than 5 percent of the total volume). When nacre fractures, the membrane is spun into silklike fibers across the gaps between the platelets, and these are probably capable of taking high loads. The interest in nacre as a material springs from the fact that it is simple in

▶ Sea urchins are made of chalk, yet can withstand the fiercest storms on a coral reef. In addition, their self-sharpening teeth are composed of some of the hardest materials in nature (see below).

structure, yet about 3,000 times tougher than the aragonite of which it is made. The reasons for this are not clear. Partly it is because a crack is always diverted as it goes from one layer to the next, but the increase in surface area (which is where the fracture energy goes in the conventional understanding of fracture) cannot account for an increase in toughness of this scale. Drying the nacre, and thus stopping the matrix material from diverting the cracks while preventing the silky fibers from being spun, halves the toughness. If we could make a material like nacre (and people have been trying for at least 20 years, sometimes with partial success), it would be at least 100 times tougher than any ceramic currently available and would use a tenth of the oil-based plastics that our tough materials currently use. It is probable, once again, that the chemistry is all wrong and that, instead of mixing an oil-based plastic with a water-based crystalline material, we should be using a water-based plastic and allowing the water to take part in the assembly process.

The self-sharpening teeth of sea urchins, some of the hardest materials in nature, are made of plates and fibers of magnesium carbonate (dolomite) in a matrix of calcium carbonate (calcite). Apart from being attractive to look at, they show us once again that, in nature, structure is cheaper than material, in contrast to technology, where material is cheaper than structure.

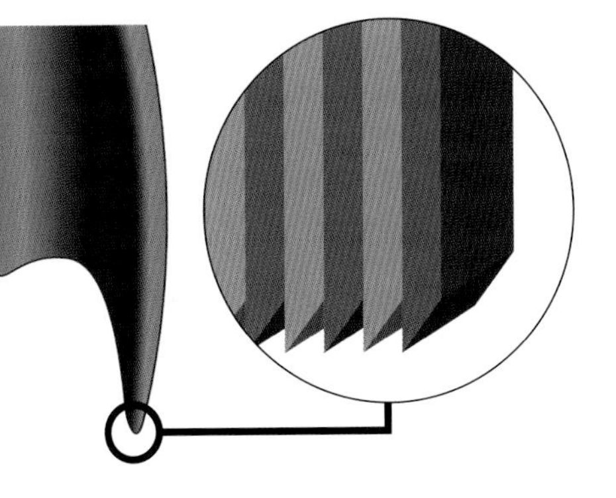

▶ The teeth of sea urchins are made of layers of soft (calcite) and hard (dolomite) materials, which wear at different rates and so generate a surface in which the abrasive structure is continually being sharpened.

Ceramics in Vertebrates

Bone is a ceramic composite that, in various shapes, supports and protects the body, allows movement, produces red and white blood cells, stores minerals, and makes antlers. Bone is hard, strong, stiff, and tough, and can adapt itself to the loads applied. Teeth are made of similar material.

About a third of bone is collagen fibers, into which nanometer-sized platelets of hydroxyapatite (a water-containing form of calcium phosphate) have been deposited, making up a further 50 percent. This composite collagen/apatite fiber is about 50 nanometers in diameter. The rest of the bone is made up of connective tissue and cells. The organization of bone is difficult to study because it is so dense. One can see the individual platelets by looking at a growth area where mineralization is only just starting, or by breaking off small wedges so that the edges are very thin and made of only one or two layers of platelets. The platelets are oriented with their crystalline c-axis along the length of the collagen fibers, but rotated along the length of the

▼ **Deer antler is made of special, very tough bone, while the neck muscles of rutting males absorb the forces of fighting. The primary bone fibers can be seen under an electron microscope (inset).**

▶ It is more difficult to get into an egg than out of it—fortunately for the chick!

fiber as well. Wet antler bone, fractured, shows individual collagen fibers that are about 200 nanometers in diameter. Hence, although the collagen-mineral composite fiber appears to be the basis of bone structure, we cannot say anything definite about its size. The self-assembly of the collagen and its shape can be affected by specific reactions with associated protein-polysaccharide hybrids. So the fiber size difference does not have to reside in the collagen.

Since bone is made up of fibers, the structure is commonly strongly oriented. This might be a disadvantage because it is unlikely, in the rigors of everyday life, that all applied forces will be nicely oriented along the fibers. A variety of structures can ameliorate this effect—the collagen fibers can be woven, made up as flat plies in sheets, or concentrically arranged, or occur in mixtures of these morphologies.

UNBREAKABLE BONE

The toughness of antler bone has been well investigated and is related to the relatively low degree of mineralization (63 percent), which results in a relatively low stiffness of about 10 GPa. Bird bone has more mineral (70 percent) and so a higher stiffness of 25 GPa. Bird bone represents a classic problem in optimization—how to make a bone that is light and stiff for flight, yet that will not break when the bird flies into an obstacle. Emma-Jane O'Leary, at Reading University, UK, investigated toughening mechanisms in chicken bone as part of a study into breakage of bones in egg-producing birds. She suggested that the yielding properties of bone could be maximized by layers of low mineralization acting as delamination lines. This is an example of the mechanism proposed for the development of aircraft structures. The degree of mineralization suggests that the toughening mechanism is necessarily structural. This fits in with the concept of microcracking. Microcracks can be only 0.02 inch (5 mm long), and they can absorb huge amounts of fracture energy as they form. They accumulate without reducing the strength of the bone, but the bone becomes more compliant. Their formation makes the bone tough because it can deform further under a given force (there are fewer of them in brittle bone). The challenge for bone construction is not so much how the crack starts, however, but how to stop it. The cracks have to absorb strain energy, but must not grow into larger cracks and cause the bone to fall apart. So crack-stopping is a crucial part of the mechanism. Microcracking is associated with clicks or acoustic events that provide the basis of an experimental monitoring technique to allow an insight into this process. The number of acoustic events increases as the bone becomes more compliant, so the two effects are closely related. Antler bone manages to keep the microcracks more dispersed and separate, and the macrocracks are tortuous, leading to increased strength and ability to resist fracture. This would, of course, be needed in the rutting season. So even the fracture of bone follows a series of hierarchical events, dictated by the hierarchical organization of the bone itself.

Recycling and Hierarchy

The plastic found in plastic bottles can be reused to make things such as my sweater. But what then? French scientist and philosopher René Descartes (1596–1650) said, "I think, therefore I am." I say, "I rot, therefore I was ... and will be!" Biology is the best recycler, and we have much to copy and learn.

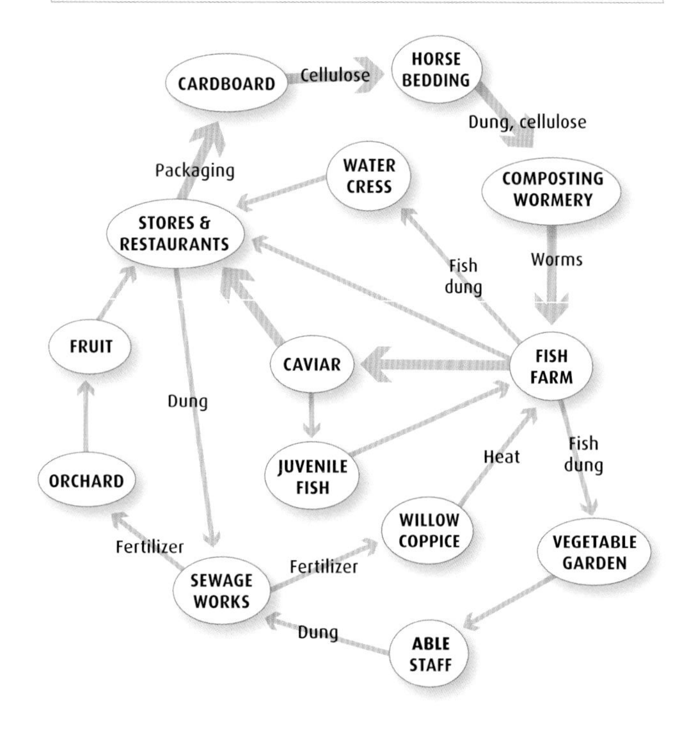

It is sometimes possible to use the biological route to recycling. This is a distinct advantage of using biological materials (or their relatives) in technical processes because the problems of the chemistry have already been solved by nature. For instance, in the "Cardboard to Caviar" project (properly known as the ABLE project), set up in the UK in 2002, cardboard collected from local businesses is shredded for animal bedding, where it acquires dung and hair. Worms in composting beds break down this bedding mixture. Sturgeons in fish tanks eat the excess worms from the composting beds. The fish are then harvested and sold as food, with some being raised to maturity for the production of caviar. The more levels through which the asset has to pass on its way down the chain, the more people can make use of the embedded energy and the greater is the perceived value.

Unfortunately, very few current schemes use the normal processes of biology to aid recycling of materials in a constructive manner, largely because our materials are rendered biologically inert through the introduction of high-energy bonds (necessarily using high temperatures). Biological materials have evolved to be recycled, and their molecules are stabilized by bonds that are only just strong enough for the expected conditions of temperature and mechanical function. Thus the proteins of most animals

◀ Recycling can be implemented to profit all. Close examination of biological recycling reveals that it is dung that makes the world go around, not love! This diagram shows how recycling cardboard undergoes various stages before being used to help produce caviar.

◀ Plumes of hot sulfides streaming from a midocean vent.

start to show signs of breaking down at 113° F (45° C), and only those of the animals living around the midocean vents can withstand temperatures in the hundreds of degrees Celsius. This means less energy is required to break the materials down during processes of digestion and more energy is available for other processes (in most organisms those "other processes" are largely feeding and sex).

HIERARCHY OF MATERIALS AND STRUCTURES

Biological materials (and structures) are hierarchical, in that they are self-assembled from the molecular level upward, which is a basic outcome of the developmental pathways of biological systems. In a biological system, the only forces available are intermolecular. By comparison with those used in technical materials processing, these forces are very weak and very short range. The common question posed by the engineer, "What is the role of structural hierarchy?", is therefore ill-formed and irrelevant. Hierarchy does not play a role, but rather it is the only way in which biological organisms can make large structures, and is therefore intrinsic. More apposite questions would be "What are the mechanisms of assembly at the various levels of hierarchy in the generation of a biological structure?" and "What selective advantage do these mechanisms deliver for the organism?"

Hierarchical construction has far-reaching consequences for the properties of biological materials and structures. Stiffness is relatively independent of the size of the individual components, relying more on the relative amounts of the components such as fibers and ceramic crystals and their interactions. Resistance to fracture, however, especially in a stiff material, depends strongly on size and shape, and it is here that the size relationships and interfaces of hierarchical construction become significant. Areas or layers that are softer than the rest can greatly affect the failure properties by stopping cracks or by diverting them. To understand this mechanism, you need to know that ahead of every crack there is a small force at right angles to the main one that is opening the crack. Thus if there is a weakness ahead of the crack, this will tend to be opened and the main crack will run into it. As a result, the sharp end of the crack will be blunted and the high localized forces dissipated and made safe. This phenomenon is taken much further in biological materials.

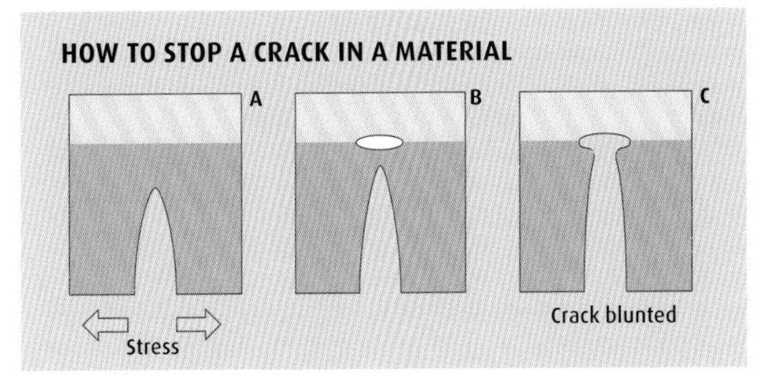

▶ The material in (A) is being stretched horizontally with a crack developing from below toward a line of weaker material. As the crack approaches the weaker layer a crack opens up ahead of it (B), which blunts the main crack and stops it (C).

Hierarchy in Small Things

Since the size of a typical crack-starting flaw in a ceramic or crystalline material is of the order of 10–30 nanometers long, it follows that, if the reinforcing particles in a biological ceramic composite are smaller than this, they will be free of flaws and will not fail in a brittle manner.

Probably the most serious problem for an engineer is something breaking. All breaks start from minute flaws in the material; in a brittle material, the flaws can be very small indeed. The brittle hydroxyapatite platelets that stiffen bone are even smaller—a few nanometers thick—so they are not susceptible to dangerous flaws at normal loads. This means that in bone stiffening can be provided by a "brittle" material without risking failure of the bone as a whole.

A flaw has to be several micrometers long before it can start a crack that will travel across a bone. This is far larger than the collagen/apatite fibers of bone; in fact, larger than the individual layers of bone that the fibers form, making the next level of the hierarchical structure. So resistance to breakage is controlled at a different size level to the resistance to bending—the stiffness—which lets them be separately

▼ The deer antler bone (below right) is clearly made from fibers (collagen-plus-mineral), while in the cow body bone (below left) the fibers are stuck together by extra minerals, which clog up the interfaces so that a crack goes straight through.

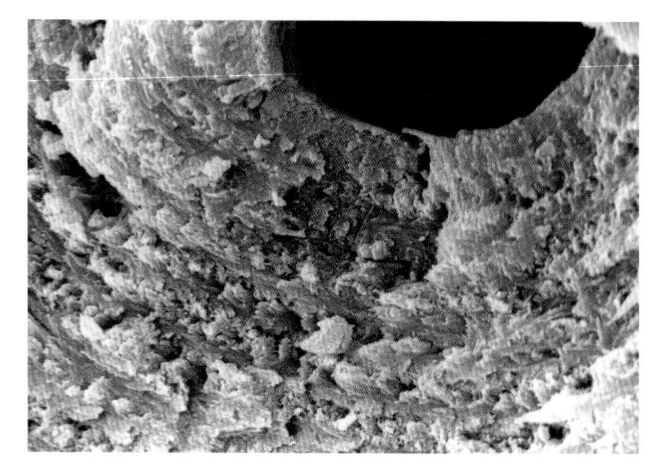

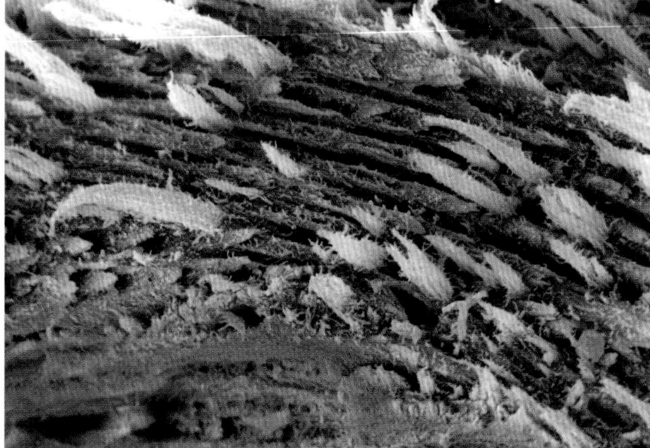

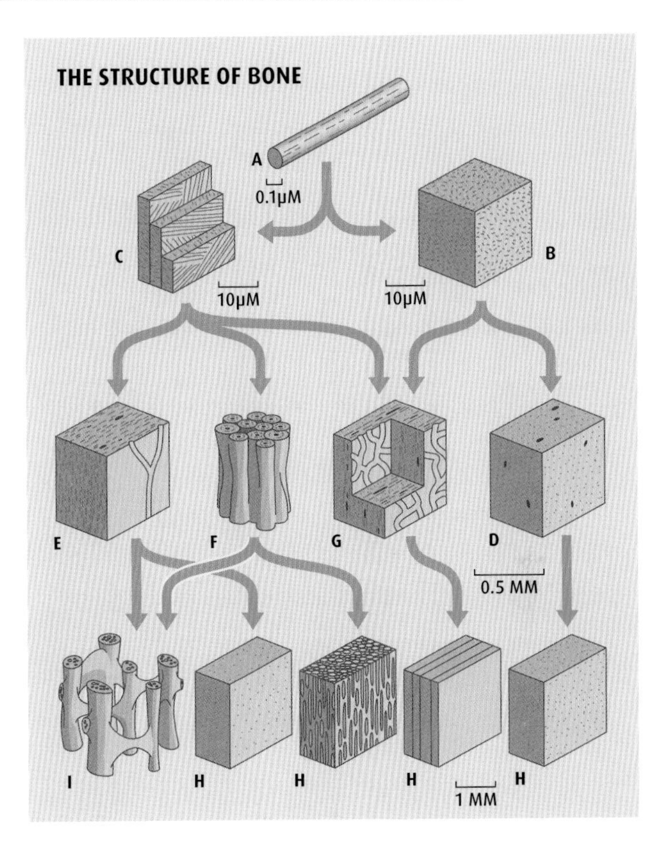

▶ Hierarchy of bone, a mix 'n' match material. (A) collagen with apatite crystals. This can make randomly arranged woven bone (B) or oriented layers of lamellar bone (C). These occur in woven bone (D), primary lamellar bone (E), osteate bone (F), and laminar bone (G). The black dots are sections through blood vessels. These types are further mixed to give various sorts of compact or solid bone (H) and trabecular bone (I), which fill the space inside the major bones.

controlled within the material. This is impossible in most human-made materials, such as metals, which are uniform in structure from the nanometer to the centimeter scale, and which sacrifice stiffness for toughness, or vice versa.

A single material—"bone"—can display a wide range of properties, depending on the relative proportions of the collagen (which contains water—never forget the water!) and hydroxyapatite, and the nature of the interfaces between the levels of hierarchy. Bovine bone, for instance, has a high proportion of hydroxyapatite; the fibers are glued together with other calcium salts. By contrast, antler bone has less hydroxyapatite, the fibers are not glued together with other calcium salts, and so it fractures differently. Whereas the fracture surface of bovine leg bone reflects little of the underlying structure because the crack travels straight through the material, the fracture surface of antler bone (a much tougher material) shows the separated collagen/apatite fibers that are the primary composite fiber of all bone. The interface at this level of hierarchy is softer in antler than in bovine leg bone, and this stops the cracks as they run into the softer material. The nature and type of damage in bone is thus also hierarchical, meaning that biological ceramics can be nearly as stiff as technical ceramics, but more durable.

PARTICLE SIZE AND HIGH-PERFORMANCE MATERIALS

Could we produce stiffer and more durable composite materials by controlling the size of the stiffening particles? The idea that nanoparticles could provide high-performance biomimetic materials is well established, but in practice it is poorly developed. Few materials actually use nanoparticles. Mostly the particles measure hundreds of nanometers in any dimension; however, if the stiffening particles are ceramic, they must have dimensions of the order of tens of nanometers at the most if they are not to break. In addition, the particles in human-made ceramic composites are not oriented as they are in bone, and bone has ten times more ceramic than we can get into a composite. Most importantly, the advantages of hierarchy—greater versatility in production and properties, in particular, the separate manipulation of toughness and stiffness—are not realized.

Much nanotechnology relies on self-assembly, yet self-assembly is really available only at the very lowest size level—that of individual molecules or aggregations of molecules. Above this size there are assembly techniques such as electrospinning and the wide range of fiber-handling techniques developed in the papermaking and textile industries. These techniques enable the production of textures such as felt and rope with all the control of interfacial properties that is found in biological materials. Along with that go the control of fracture and the control and development of toughness and durability.

Hierarchy in Larger Things

The study of biological materials shows that the introduction of heterogeneity and interfaces, at the right size levels, can improve durability without unduly affecting load bearing. So perhaps all one needs to do is to introduce heterogeneity over a range of size scales.

One of the aims of materials processing in technology is to produce a uniform material, on the assumption that this will make its properties predictable. Iron, however, is a versatile material precisely because it can be heterogeneous (think of all the different types of steel, each with a range of specialized properties), with layers produced by the smith's working, and atoms from other elements alloyed between the crystals causing important differences and improvements in mechanical performance.

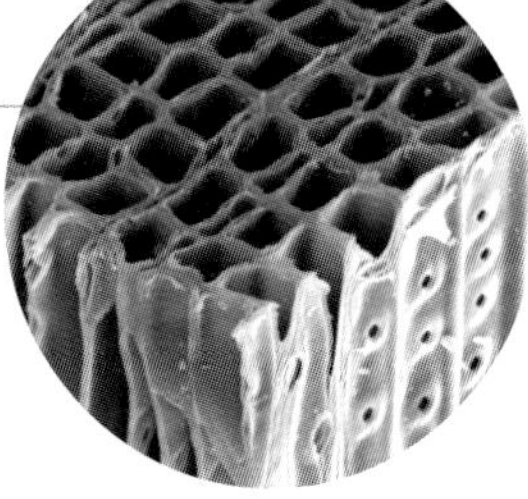

▲ The main cells in pine wood are about 20 nm in diameter and can be a millimeter or more long.

HIERARCHY FOR STRENGTH

Hierarchy is also evident in larger organic structures. For example, in wood the vessels that transmit water up the trunk of a broad-leaved tree are much larger (500 μm diameter) than most of the wood cells (50 μm diameter). When the wood is compressed across the grain, these large vessels collapse first and in doing so entrain the collapse of surrounding cells. Depending on the distribution of the vessels, the wood can be brittle (for example, oak, where the vessels develop in a single layer in the wood—"ring porous"—concentrating collapse into a small zone) or tough (for

▼ Hierarchy of collagen, going from molecule to tendon.

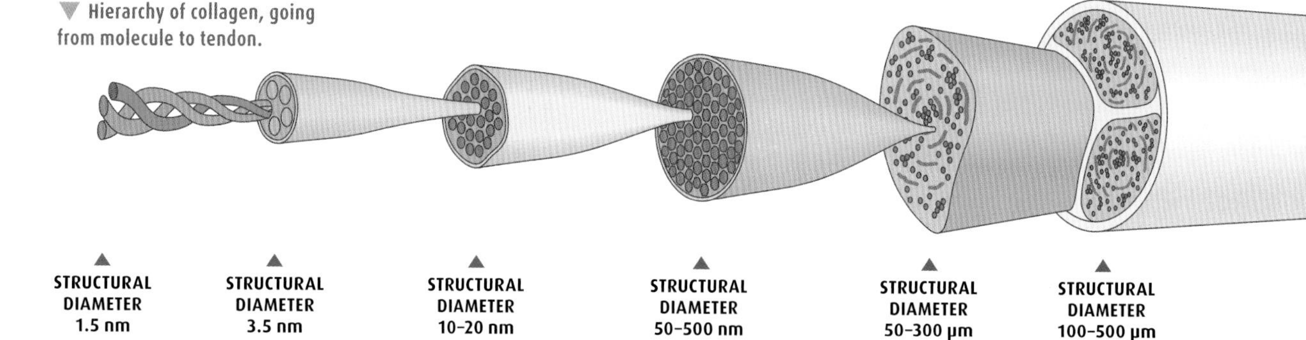

◀ You cannot propagate a crack through a grass leaf—so a grazing cow just holds on with its tongue and pulls.

example, hickory or beech, where the vessels are distributed evenly throughout the wood—"diffuse porous"—and so spread the collapse among many more cells). The presence of these large vessels enhances the longitudinal properties of the wood as well. Professor Roderic Lakes of the University of Wisconsin reported making hierarchical honeycomb by stacking unexpanded small cell layers of honeycomb with wide stripes of glue between them, so that when expanded they became cells whose walls were made of larger cells. This structure looks much like a piece of hardwood. He found that the compressive strength of the hierarchical honeycomb was about 3.5 times greater than that of the original honeycomb.

An example of hierarchy in engineering materials/structures is rope, which is made of swathes of twisted polymeric or metal fibers laid up against each other in increasingly complex units. Ropes are very tough partly because the fibers are kept separate and it is difficult for a crack to travel across the structure. Rope structures in expensive high-performance tires are more flexible and durable. Biological equivalents of rope, where the fibers are separated, are found in lianas and grasses (it is no coincidence that ropes were originally made from plant fibers). Grass is so difficult to break that large animals such as cattle simply grab hold of a tuft with their tongues and pull. Smaller animals such as insects have developed scissorlike mouthparts to cut through the fibers.

HIERARCHY FOR ADAPTABILITY

Hierarchy is a direct outcome of self-assembly in biology, itself driven by information from molecular ordering. The versatility of a material is enriched by more structuring at each level of hierarchy, so that adaptability increases with the number of levels. There is even more adaptability available with biomimetics. Biology, saving energy, uses a limited range of chemicals, but we can increase that range and reap the benefits. For instance, there are many more amino acids available than the 20 or so in biology, so we could produce a wider range of proteins while still keeping the advantages of biological systems. Better, we can use outside sources of energy to control temperatures and speeds of reaction; we can use machines to organize molecules into fibers and complex shapes, producing materials with no waste.

▲ Fibers are twisted together to form bigger fibers, which in turn are twisted together to form yet bigger fibers, which are twisted together to form ... eventually, rope.

How to Make Small Structures

To make microelectronic devices, a layer of material is first deposited. An image or pattern made of resistant material is deposited onto this layer and the exposed areas dissolved away or etched into the surface. This process is repeated a number of times to make very sophisticated structures.

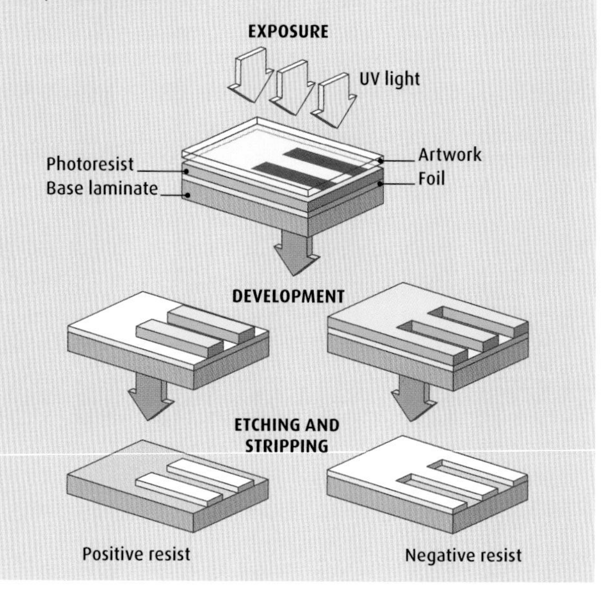

PHOTOLITHOGRAPHY
The principle of photolithography, repeated many times, will yield complex structures.

Although this technology is very impressive, it is eclipsed by the generation of shapes in natural materials that occurs at the molecular level, with controlled adhesion and mechanical forces developed by surrounding cells. This is at least two orders of magnitude smaller than current technology can manage. One of the most intriguing shapes is the branching structure of the "adhesive" hairs on the foot of the gecko. They are far finer than anything generated by etching processes, being of nanometric dimension at the tipmost branches. Detailed investigation into how these hairs are formed could prove enormously useful. The amazing adhesive properties of the gecko's foot are covered in more detail in Chapter 1.

BOTTOM UP IS BETTER!

Similar structures can be generated by droplets of liquid falling through another liquid of lower viscosity, a process that could produce hairs of molecular dimensions at the tip, and far more cheaply and quickly. There is plenty of opportunity for a freer approach to materials processing, using physical processes based on surface interactions and surface energy. As an example, in the developing tendon of a 14-day-old chick embryo, there are at least three types of compartments formed on the outside of the cells by folding and intucking of the cell wall. The first consists of a series of narrow channels, about 150 nm in diameter, originating deep within the body of the cell and containing at most two

▶ Organisms such as this chick embryo grow by self-assembly, driven by intermolecular forces at small sizes, and by molding and external forces such as gravity at larger sizes.

or three fibrils. These channels fuse laterally to give rise to the second type of compartment, 2–3 μm across and defined by single or adjacent fibroblasts, where the fibrils fuse into fiber bundles. The third type of space is larger still, formed cooperatively by two or three adjacent fibroblasts, in which the fiber bundles start to look more like tendon. This compartmentalization seems to be a general phenomenon operating in different tissues to regulate the assembly of collagenous tissues. Depending on the type of tendon, lateral fusion of fibrillar elements continues after birth as a function of the animal's general growth and maturation. This system could be turned into a production line like that used for spinning yarns, starting with a fiber that is built up in stages by specific processes, except that the fibers come with their own chemical assembly mechanisms built in. The concept is of assisted self-assembly, which supports longer-range interactions and is rather like molding.

Another example of assisted self-assembly occurs in the production of the protein threads which anchor mussels to rocks (see page 136), where a range of protein types are secreted into a groove in the "foot" of the mussel. This soft mold in some way ensures the orientation of the fibrous proteins along the length of the thread and assists its self-assembly. The chrysalis performs the same function for the moth that is developing inside it. The shapes of the legs, wings, and other structures can be seen on the outside of the chrysalis, directly molding the developing insect. Such soft templating methods, creating environments for the assembly of molecules, greatly simplify and speed up the generation of materials at the small end of a hierarchical series. This would silence the critics who say that biology works too slowly to be industrially useful.

◀ The gecko's footpad is covered in branched "hairs," a few nanometers at the tip with flat ends, which rest closely to the surface and hold on by very weak forces.

Biomimetic Skins

Biomimetic skins are soft. Membranes for medical use ("scaffolds") have an open texture and support a water-retaining gel that in turn supports cells which can make a proper skin. The other main area of interest lies in surface structures modeled on leaf and insect surfaces producing breathable self-cleaning materials.

Biomimetic skins have not received much attention, but biomimetic surfaces have become very important and yield an interesting example of the way in which biomimetics currently works. One example is superhydrophobicity and the development of self-cleaning surfaces based on the "lotus effect" observed when water droplets form on lotus leaves. The concept is a product of two effects: a hydrophobic surface created by a layer of hydrocarbons, usually crystalline, and a surface texture of bumps at a spacing of about 10 μm.

◀ The water-strider has a water-repellent plastron of short hairs on its legs and so can be supported by surface tension.

The insect plastron, an arrangement of "hairs" a few micrometers long arranged at a density of 10^7 or more per cm^2, comes into the same general category of highly textured surfaces. The plastron is a well-researched system for repelling water, keeping a thin layer of air next to the surface of the animal and providing an air–water interface at the tips of the hairs across which exchange of gases can occur, thus giving the animal a form of gill that enables it to breathe underwater. This is not only a common adaptation of insects living in, on, and around water, but also obviously a technique for keeping the immediate skin of the animal dry (since water cannot penetrate between the hairs) and insulated from the cold.

A textile similar to velvet has been developed by a UK sportswear company called Finisterre as a breathable and warm waterproof material for surfers to wear. The textile's inventor, Tom Podkolinski, did not realize that he was reinventing the plastron, although he based his initial concepts on the waterproof pelage of seals and otters. Another technical version of the same concept is provided by the surface of an extremely water-repellent foam that mimics this mechanism and allows direct extraction of oxygen from aerated water. Water-repellent surfaces are also being developed to make devices that can walk on water, using design ideas from *Gerris* spp., the water-strider (see above, opposite). More on these robotic devices can be learned from Chapter 1.

◀ Microscopic view of the surface of a lotus leaf. The waxy bumps, invisible to the naked eye, are covered in small wax crystals that prevent particles from sticking to the surface.

◀ The lotus plant (*Nelumbo nucifera*) grows in swampy water, yet emerges from the dark depths of the swamp without a blemish. The rain washes the surface of the plant clean.

UNDERSTANDING THE LOTUS EFFECT

Normal Effect

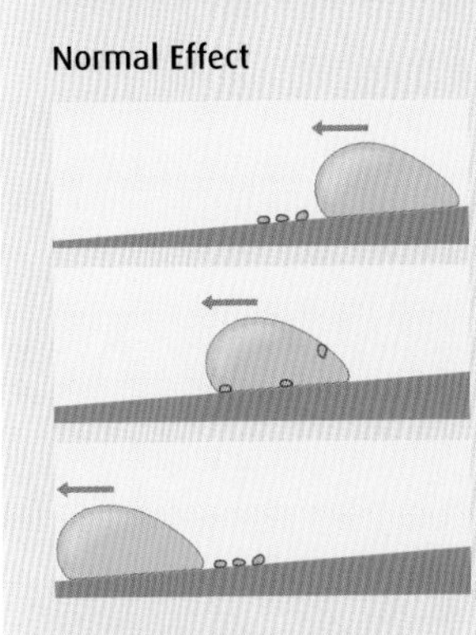

A water droplet falls on a smooth surface and forms a hemisphere with little momentum.

The droplet slides over tiny dirt particles, but cannot pick them up.

The water moves on, but the dirt particles are left behind. The drying surface therefore remains dirty.

Lotus Effect

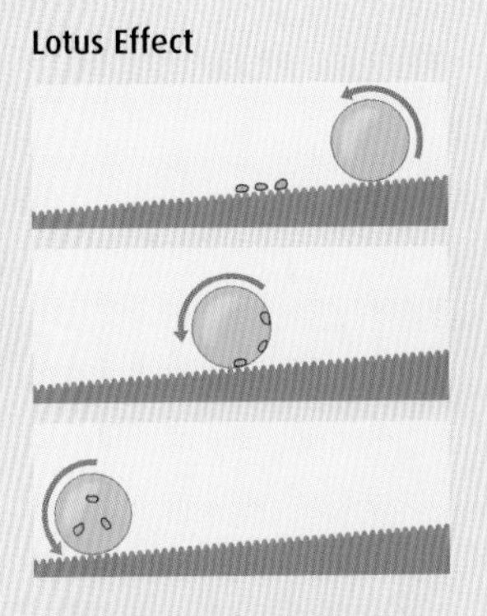

On a rough surface, the droplet remains almost spherical because it cannot settle.

It bowls along at speed, picking up dirt particles. It has a greater surface area for them to stick to.

The droplet rolls off, carrying the dirt particles with it and leaving the surface clean.

Dirt particles will not stick to a surface painted with StoLotusan® (above), a water-repellent exterior coating, so the rain washes the walls clean.

Biomimetics in Application

Biomimetics has made limited inroads into technology. The major areas of interest are robotics (including sensors, actuators, and locomotion), color (mostly physical color, including photonics), materials (mostly tough ceramics and fibers), and material surfaces (including textiles).

Biology has always provided inspiration for architects and artists of all sorts, but the design of hierarchical structures, which has served its time in the Eiffel Tower (which was designed to last for 20 years and has lasted more than six times longer), and reached its zenith in the successful R100 airship, has hardly been explored. The

▶ The innovative R100 airship was a hierarchical structure, so it was very light and strong.

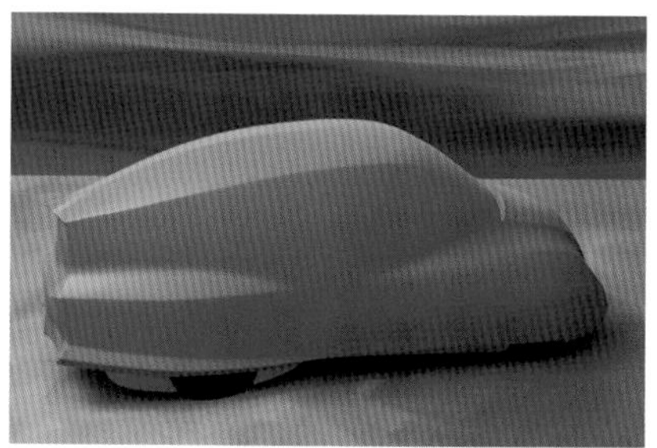

Eiffel Tower is ten times more efficient as a structure in terms of strength, and in economical use of materials, than the nearby Pompidou Center, with the corollary that a building can be made with a tenth of the structural material if it is designed using principles of hierarchy. Such a building would obviously require not only less material for its manufacture, but also less robust foundations, and it would be cheaper to make, although probably more expensive to design. The two structures are, however, designed for different purposes, and so extrapolation of this approach in general has to be taken in context.

The automotive industry is using more biological materials, such as plant fibers. Parts of the recent and much-touted boxfish concept car, a design study by Daimler-Chrysler, used a stress-relieving design tool developed by Claus Mattheck, who studied the growth of trees to gain an insight into how nature would deal with the strength-to-weight problem. The resulting weight saving was of the order of 40 percent, but the process was discarded by the car producers with the comment that it took too much time in fabrication. The main historical area of biological influence on technology is flight, on which there is extensive literature comparing animals and aircraft. But in general, biomimetics, either as direct design tool or as some sort of abstraction, has made few inroads into technology.

THE BIOMIMETIC PARADIGM

There is one factor in the relationship between biology and technology that we have not mentioned. Technology is never questioned as to its completeness or quality. That is not to say that technology is not continually improving, but the tacit assumption is that technology and the way we engineer things represent the best practice. Current environmental problems scream at us that this is an untenable assumption; the ways of biology are benign, at least upon this planet. Yet the current practice in biomimetics is to incorporate mechanisms and structures found in biology into the current engineering paradigm. Is this the best we can do? Of course, it is impossible to interrogate technology from within the system. Should we not be examining the ways in which biology delivers functions (the same functions that engineering delivers) and see if there is a better way of doing it? Might this not be the major advantage that biomimetics can deliver? We need ways of systematizing technology and biology.

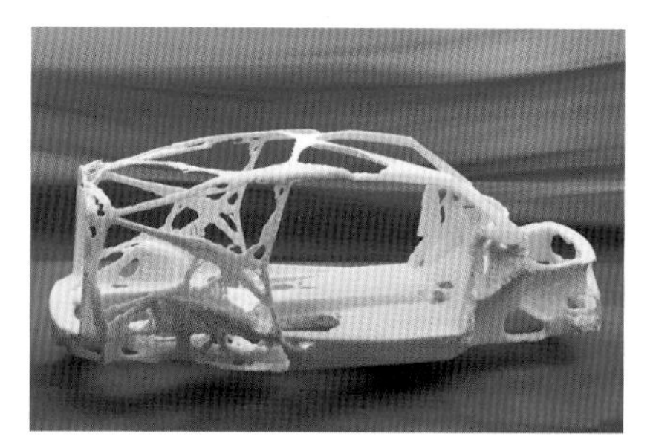

▲ The nave roof in the Sagrada Família (designed by Antoni Gaudí, but only now being built) is a motif of leaves supported on plant stems.

◀ The DaimlerChrysler design concept, the boxfish car: streamlined like a fish, chassis designed like a tree.

Technology Transfer

The transfer of a concept or mechanism from living to nonliving systems is not a straightforward matter. Currently there are three different levels for transfer: direct copying; connecting ideas using word chains; and understanding functions using standard methods for solving problems.

The first and most common route is a simple and direct replica of the biological prototype. The technical abstraction is possible only because a biologist has pointed out an interesting or unusual phenomenon and has uncovered the general principles behind its functioning (e.g. the self-cleaning lotus effect). Only then does the biological principle become available outside biology for biomimetic use. The result is often unexpected (e.g. waterproof breathable clothing). The advantage of this approach is that biology has already done a lot of the design prototyping; the disadvantage is that every example has to be approached afresh and that it is very difficult to transfer concepts from one project to the next. The outcome may be novel, but, since it obviously required input from physics and engineering to understand the significance of the function of the biological prototype, there is little likelihood of fresh thinking. Indeed, the main way in which a novel biological mechanism can be identified is that it has been observed and analyzed in an inert system that is simpler than the biological one. One can only see what one knows and expects. It is likely that the most important and useful ideas to come out of biomimetics will be the most difficult to identify because they will have little or no parallel in the technical world. An example is fracture. Few people realized how important the fracture toughness mechanisms of biological materials were because biological materials so rarely break. I know from my own experience that this makes it very difficult for a biologist to appreciate that fracture is an important topic area.

BIOLOGICAL KEYWORDS

A second route for transfer that is more abstract (and therefore of more general use) is based on vocabulary and meaning. Lily Shu, a design engineer at the University of Toronto, mapped analogies by analysis of natural language. She identified biological keywords suggested from the problem to be solved in engineering, then organized and ranked them. She has done this for the engineering functions of cleaning, encapsulation, and microassembly. Her search starts with a digital version of biology gleaned from textbooks and research papers. The keywords to be used are specifically verbs because they imply functionality; the starting point in engineering is derived by listing the functions that are to be delivered. The problems arise when the engineering words are compared with those in biology. Shu uses the example of cleaning in engineering, which has hardly any equivalent in biology. But the question "What is the function of cleaning?" yields the answer "to combat contamination," which in turn elicits the biological equivalent of defense against pathogens. This becomes

PERSISTENT URBAN MYTHS AND MISUNDERSTANDINGS

The hairs of polar bear fur are supposed to act as light guides, bringing heat to the skin surface. Experimentation shows this to be impossible because the hair has been found not to transmit light after all.

Victoria amazonica leaves were supposedly the inspiration for the greenhouse roof designed by Joseph Paxton (1803–65) and the roof of the Crystal Palace in London, UK, but the geometry and structural principles of the underside of the leaf (shown here) and that of the roofs are totally different.

The Eiffel Tower was not modeled on the structure of the human femur or a tulip stem, but it was one of the first buildings to take into account wind-loading effects, so that bending forces are reduced and evenly fed into the foundations.

a fruitful path: defend → invade → evacuate → eliminate → remove → clean. The next stage is to see how organisms defend themselves against invading pathogens, using antibodies, leukocytes, encapsulation, and isolation. In order to make this process computable, the linguistic environment of the word is also analyzed, so that its position in a semantic hierarchy becomes a part of its description and defines the context. This context then provides extra keywords that help to define the bridging concept being sought. The importance of the various bridging words is then reflected by their frequency of occurrence, which can be used to guide the engineer toward more relevant or more useful biological examples of the phenomenon being sought for production of analogy. Thus these techniques suggest more useful analogies for technology.

Obviously this bridging process, being linguistic, is not limited to biology, and can be used for any domain, given the availability of appropriate domain-specific knowledge sources and references. A main advantage of this approach is that, in its final form as a tool, it appears not to need any biological input and so can be performed by an engineer or other domain specialist.

This method, however, still requires a proper definition of the problem (not always an easy task) and relies upon the quality and completeness of the biological text for its reliability. A catalog of the biological domain combined with lexical analysis can show the nonbiologist which areas of biology are likely to yield information on functions relevant to the problem to be solved. But neither the catalog nor lexical analysis properly underpins biomimetics. Neither recognizes that the number of manipulations available for changing a system (for example, so that it will produce the function required) is likely to be finite, although there is the assumption basic to biomimetics that the manipulations developed within biology can be translated to engineering. It seems obvious that rather than trying to develop a technique *de novo*, one should adapt one that has already been developed.

Directed Problem Solving—Theory

The third method of knowledge transfer says that all major functions that contribute to the success of humankind have been discovered and recorded, since they rely on physical principles, and engineering's advancement represents the development of more advanced ways of delivering those functions.

Most people, when faced with a problem, will proceed to "solve" it with an object, but a much more innovative approach is to identify carefully what your ideal result would be in terms of a delivered function, then to work out what you need to do to get to that ideal result.

TRIZ

An Uzbek engineer and researcher, Genrich Altshuller, developed a rigorous set of methods that concentrate on defining the function to be delivered (hence independent of the hardware and capable of much more basic generalization), the working environment or context in which it is to be delivered, and the resources available within that environment. Frequently he found that the solution is best delivered by altering the environment, rather than by altering the immediate mechanism of delivery (compare this with the biological insight that conservation by the maintenance of the environment within which animals and plants exist is far more powerful than attempts to conserve the individual species within that environment). He called this set of methods "The Theory of Inventive Problem Solving" (in Russian, *Teoriya Resheniya Izobreatatelskikh Zadatch*, or TRIZ).

A major insight was that the manipulations required to deliver these ideal results could be summarized and reduced to a manageable number (about 40, known as the "Inventive Principles"), and that these manipulations were relevant across a wide field of applications. Altshuller also showed that any branch of technology could provide a solution to a specific problem, even if that branch does not

▲ Genrich Altshuller (1926–98), the originator of the TRIZ system of problem solving and creativity.

▲ The German philosopher Hegel (1770–1831), whose philosophical thinking forms part of the basis for modern problem-solving techniques.

▲ Heraclitus (c.535–c.475 BCE) defined a problem as being composed of a wish, an obstacle to its fulfilment, resolved by a solution.

appear to be related to the problem or its apparent solution. This is because if you define an innovation in terms of the function it is desired to deliver, and express it in simple and relatively abstract terms, you can destroy the barriers that most people erect between the various areas of their knowledge. The temptation is to define the innovation in terms of a technology ("What you need is a . . . "). Thus we are able to reinforce creativity and technology transfer by accessing ideas outside our self-defined (and thereby self-limiting) area of competence.

To add a fourth statement to a well-known categorization of knowledge, made famous of late by Donald Rumsfeld, there are unknown knowns: that is there are things that we do not know that we know (taken here to be a prime source for creative thinking). And the rules of thinking that TRIZ teaches show you how to access those unknown knowns.

FUNCTION IS THE KEY

One of the TRIZ techniques is to define a problem in terms of what function is to be delivered (e.g. greater load bearing, perhaps by a bridge that has to take more traffic or heavier trucks) and the downside of delivering that function (e.g. a heavier or more bulky structure needed to support the extra weight).

This is similar to German philosopher Georg Wilhelm Friedrich Hegel's notion of thesis and antithesis, where thesis is the statement and antithesis is the opposite of the statement. This in itself is the framework of a problem, which would not exist unless there were apparently conflicting requirements. The Hegelian resolution—the solution to the problem that somehow manages to combine thesis and antithesis—is synthesis, which in TRIZ terms is an inventive principle. These principles have been derived from examining more than three million patents, and so are a definitive collection of best practice in engineering.

From the point of view of biomimetics, if, by defining functions, we can similarly identify thesis, antithesis, and synthesis groups from biology, we can then compare the ways in which biology and engineering deliver a range of functions, and thus produce an independent assessment of whether engineering stands any chance of being environmentally benign by comparison.

Directed Problem Solving—Practice

To demonstrate the TRIZ method in action, I made a detailed analysis of how an animal might be constructed, using the example of the cuticle of an insect: eyes—transparent; jaw—very hard; teeth—even harder; leg hinge—soft, like plastic wrap; shield—thin but strong; spines—hard and sharp.

I made a list of the functions of the cuticle of insects, covering skeletal properties, design characteristics, waterproofing, and so on. Items were compared pairwise, and apparently conflicting requirements were listed. Thus the skeleton has to be stiff to provide support, but flexible to allow movement. These pairs were analyzed using TRIZ, and the derived inventive principles (representing how an engineer would solve the problem) were compared with the way in which the insect solves the problem, taken from the same list of inventive principles. Thus two lists were generated comparing technical and biological solutions to the same problem. The inventive principles were similar in only 20 percent of the thesis/antithesis pairs, suggesting that biology solves problems and delivers functions in a very different way than technology.

In technology, the most common solution to a problem is to change one of the processing parameters, for example, temperature or pressure. Most of the functions of insect cuticle, however, are provided by detailed control of properties over a very short distance, which suggests that technology should be aiming at producing not just very small components, but also integrated

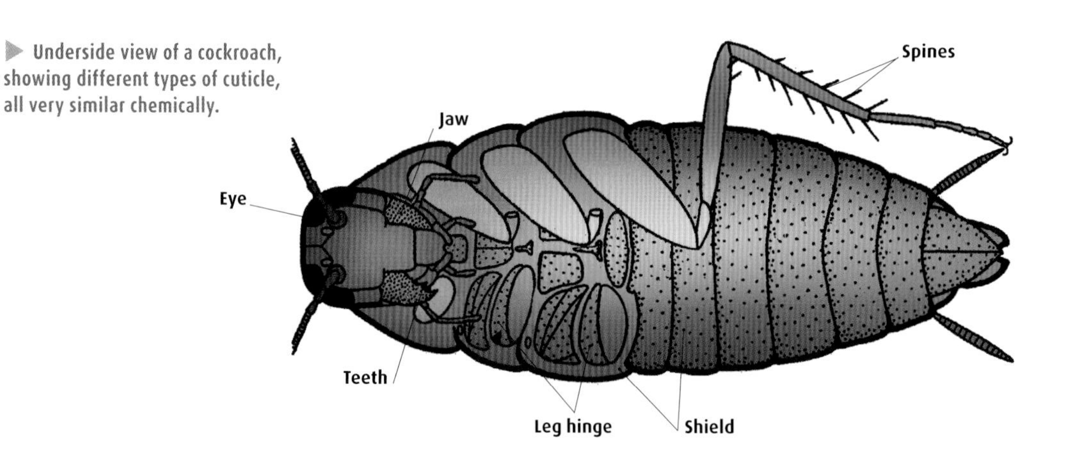

Underside view of a cockroach, showing different types of cuticle, all very similar chemically.

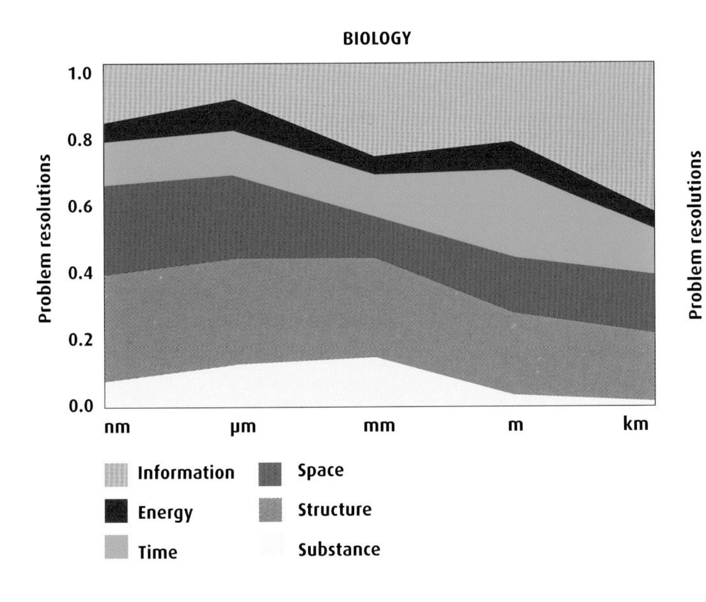

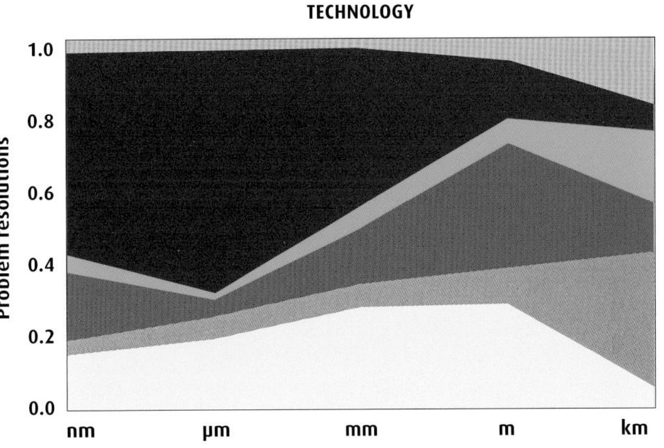

▲ Diagrams showing how six parameters, which cover all possibilities, are used to control processes and solve problems in technology and biology. Materials processing occupies the size range from nm to μm; manufacturing covers the range from mm or cm to tens of m; societies cover the rest of the size range. What a difference between them! (courtesy of The Royal Society).

assemblages of components. An example of the success of this approach can be seen in the design and integration of the campaniform sensillum into the cuticle of insects. This sensillum is a displacement sensor, relying on the deformation of a hole through the cuticle. The hole is formed by diverting the chitin fibers in the cuticle, rather than allowing the fibers to end blindly at the edge—the equivalent of drilling the hole in a sheet of fibrous composite. This attention to detail and local quality of the design results not only in greater safety (the stress concentrations usually associated with a hole are almost totally avoided), but also in an increase of an order of magnitude in the local amplification of globally applied strain, leading to increased sensitivity of the sensillum. Thus, the proper integration of the strain sensor results in significant technical advantage, relative to, say, the technological solution of sticking a foil gauge onto the surface.

BIOTRIZ

The TRIZ approach has applications within biology because it allows an objective approach to understanding of functions at all levels of complexity. We divided the modes in which we can manipulate a system into the headings of substance, structure, energy, space, time, and information. These six parameter groups cover all possibilities, and as thesis–antithesis pairs give 36 (i.e. 6 x 6) possible classifications for a problem. We took about 2,500 examples of biological functions over all size scales, from molecule to kilometer, decided which inventive principle they each represented, and classified them according to which of these parameters represented the thesis–antithesis pair. We called this matrix "BioTRIZ." We reduced the TRIZ system to the same level and compared the two matrices that resulted. This revealed only 12 percent similarity between TRIZ and BioTRIZ. At the level of materials processing, 70 percent of technological problems are solved by manipulating energy and 15 percent by manipulating materials (i.e. substance). In stark contrast, energy figures hardly at all (no more than 5 percent) in the solution of problems in biological systems. Instead, 20 percent of biological design problems are solved by manipulating the way things are put together (i.e. structure) and 15 percent are solved by manipulating the transfer of information (derived from the sequence of bases on DNA) within cells and tissues.

Is Success an Option?

BioTRIZ has not been available for long but has already had success. Salmaan Craig, an engineering student with the firm Buro Happold, looked at buildings in areas with hot climates, which can get uncomfortably warm during the day. The insulation he designed in response allows heat to be radiated back into the night sky.

Using BioTRIZ to analyze the problem, it turned out to be possible to make the insulation keep out heat from the Sun (short-wave radiation) during the day, but provide a route out for long-wave radiation from the building during the night. This was achieved by using the concrete of the roof as a heat store during the day and structuring the insulation to provide an exit pathway for radiation by giving it an orientation. This can be classed as a biomimetic solution to the problem because it was derived from the suggestion that the strongest design variable at the size level of roof insulation is "structure." The engineering

▼ **Buildings in very hot areas tend to have thick walls, which are slow to heat up in the day, but are also slow to cool down at night.**

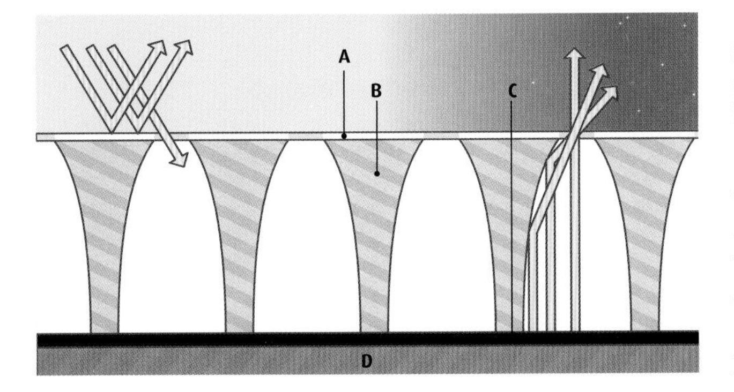

Salmaan Craig's roof, which insulates during the day (layer A reflects sunlight and layer B insulates) and radiates (via the black layer, C) heat (stored in the concrete, D) during the night. (Labels: A = reflector; B = insulator; C = radiator; D =heat store)

answer would be "energy," and indeed most people would automatically see air-conditioning as the solution to an overheated building. This would be environmentally acceptable only if the air conditioner were driven by solar energy via photoelectric cells!

Our studies have analyzed and cataloged how biology and engineering solve the same range of problems, and have presented the results in a single, uniform format. Thus we can compare biology and engineering at the same level of detail and function. Although most engineers would not readily admit it, biomimetics is the first true challenge to engineering on its own territory, and in many respects it finds engineering wanting. This challenge has some urgency about it, since worldwide our consumption of energy far exceeds the instantaneous supply. Therefore we cannot, at present, be "sustainable." But could biomimetics help us? It is disputable whether the biological system that we inherited on this planet is the paradigm for sustainability, but there is no evidence that any other system would provide us with the resources we need for survival, at least on Planet Earth. Biomimetics can point us toward a more sustainable future, if we can only transfer its lessons to our technology or, perhaps better, translate our technology into a biological format. Biomimetics can have a significant effect on our survival if we can use it to redesign a technology that it is not so reliant on energy. This is eminently possible, and with the tools currently available we can produce a wide range of materials, structures, and systems to supplement the current ones.

Many of the techniques we need for the production of materials and structures are available, some of them modern versions of traditional methods. Self-assembly and the control of interfaces remain important research areas where biology does much better than we can at present.

Glossary

Acoustics the scientific study of sound

Actuator a type of transducer that converts energy into some kind of mechanical function, e.g. a muscle or a motor

Algorithm a set of rules used to govern a process or solve a computational problem

Amplitude the height of a wave, sometimes measured from the midpoint between oscillations to the peak of the wave and sometimes from the peak to the corresponding trough

Anisotropy a property of a substance (e.g. refractive index, absorbance, elasticity) that varies depending on direction. Timber displays anisotropy in that it is easier to split in the direction of the grain than across the grain

Arthropod an invertebrate animal having a segmented body, jointed limbs, and an exoskeleton. Insects, crustacea, and arachnids are all arthropods

Atmosphere a unit of pressure, defined as 101.325 kPa, more or less the mean atmospheric pressure at sea level

Attenuate to weaken, used particularly of light passing through a dispersing medium such as water, or of sound waves losing intensity when traveling through an absorbing medium

Bioinspiration using inspiration from nature as a basis for developing engineering solutions to problems

Biomass the total quantity of living biological organisms in a given area

Biomimetics mimicking nature in some way to develop engineering solutions to problems

Birefringence the decomposition of a ray of light into two rays when it passes through a substance, a common property of crystals, which have two different refractive indices

Boundary layer the layer of fluid that lies closest to the bounding surface, e.g. the seabed, a pipe wall, or an aircraft wing

Brittleness a property of materials that are liable to fracture under stress without any prior deformation. A material can be both strong and brittle at the same time

Broadband relating to a wide range of frequencies with reference to either a signal or a receiver

Calcification the formation of calcium compounds, an important process for the development of shells and bony structures

Capillary wave a wave traveling along the surface of a fluid, otherwise known as a "ripple"

Case-based reasoning the process of solving a problem on the basis of similar previous problems

Cavitation the formation of bubbles in a liquid by the movement of an object through it; the bubbles occur in very low pressures in turbulence, resulting in bubbles that can, for example, damage marine propellors

C-axis the principal axis of a structure

Cellulose a polysaccharide consisting of a linear chain of many hundreds of sugar molecules. Cellulose is the structural component of the primary cell wall of green plants and many forms of algae

Central pattern generator (CPG) a neural network that is able to act independently of the central nervous system of an animal to generate rhythmic outputs (e.g. movements such as walking, breathing, or flying that are repetitive). CPGs can generate these outputs without a corresponding external rhythmic input, e.g. walking occurs without a separate message from the cortex for each step

Ceramic an inorganic crystalline mineral. Ceramics are characteristically hard and brittle, and naturally occurring ceramics are important elements in skeletal structures

Chitin a tough, protective, semitransparent substance, primarily a nitrogen-containing polysaccharide, forming the principal component of arthropod exoskeletons and also found in the cell walls of certain fungi

Chromatophores pigment-containing and light-reflecting cells found in cold-blooded animals such as fish and reptiles that produce skin color

Cilia minute hairlike vibrating structures on the surface of single-celled microorganisms that create currents in a surrounding fluid or are used for locomotion

Collagen a protein, the main structural protein of connective tissue in animals

Composite a material made from several constituent materials that themselves remain distinct and separate. Composites are most often made up of a matrix and a reinforcement material; the latter lends its specific properties to the composite, while the matrix supplies a stable environment

Control surfaces movable elements on the body of an aircraft or submarine such as elevators, ailerons, and rudders that allow the attitude and direction of the vehicle to be controlled

Convection the movement of molecules within a fluid, often associated with the transfer of heat energy. "Free" convection occurs naturally as warmer molecules become more buoyant and rise. "Forced" convection results from an external agent such as a fan or pump that causes the fluid to circulate

Covalent bond an atomic bond that involves the sharing of a pair of electrons by two atoms, the strongest chemical bond

Cross-links bonds that link one polymer chain to another. Strongly cross-linked materials are stable and difficult to break down

Crystallinity the degree of order or regularity in the internal molecular structure of a material, governing properties such as hardness and density

Cytoplasm the part of a cell that is enclosed within the cell membrane

Cytoskeleton the internal "scaffolding" of a cytoplasm

Distal situated at the region farthest away from the point of attachment, e.g. of a limb, hair, tail, etc.

Ductility a property of materials that deform under stress before fracturing, and particularly applied to metals that can be stretched into wires

Elasticity the property of a material whereby it returns to its original shape after the application of a stress that has deformed it

Elastin a protein in the body's connective tissue that allows tissues to resume their shape after stretching

Emergence the way in which complex systems develop from the interaction of a group of simple processes or events

Encapsulation the process of coating a particle to protect it from the outside environment, or to protect the outside environment from it

Enzyme a protein that speeds up a chemical reaction

Epibiont an organism that lives on the surface of another organism

Fibril a tiny fiber around 1 nanometer (nm) in diameter

Foil a plate or fin that controls lift or movement, e.g. an airfoil on an aircraft or the fins of a fish

Frequency for any repeated cyclical action, e.g. sound waves, the frequency is the number of cycles in a given period. The unit of frequency is hertz (Hz), which is a measure of the number of cycles per second

Fuzzy reasoning the process of solving a problem by considering various approximations to a solution and narrowing down the variables, a method of reasoning common in humans, but very difficult to replicate in artificial intelligence systems

Gradient the difference in a particular property—e.g. temperature, velocity, molecular concentration—between adjacent regions of a material

Group intelligence a feature of super-organisms in which many individual agents with limited intelligence and information are able to pool their resources to accomplish a goal beyond their individual capabilities

Helix a spiral curve. Many chemical substances have a helical structure

Hydrocarbons the simplest organic compounds, made up of hydrogen and carbon

Hydrodynamics the study of liquids in motion

Hydrophilic a chemical substance that readily forms bonds with water is described as "hydrophilic"

Hydrophobic a chemical substance that repels water is described as "hydrophobic"

Interface a common boundary between two regions, or a region where two or more systems interact

Keratin a fibrous protein that performs structural functions in animals

Kinetic energy energy that an object possesses as a result of its movement

Lamella a thin platelike structure

Laminated consisting or made up of layers. Lamination is a common method of improving resistance to fracture

Leukocytes white blood cells that defend the human body against infection

Lignin a polymer found in trees and woody plants that binds the cells and provides structural support

Load sharing a means of decreasing the impact of a force or stress on a system by disseminating its effects across a series of subunits

Locomotion movement from one place to another

Lubricate to reduce the friction between two surfaces, often by means of a fluid

Matrix an environment or a substance in which a structure grows or is embedded

Microcracking a mechanism in ceramics that increases their resistance to fracture by permitting the energy of a large crack to be dissipated into a series of smaller cracks and thus arrested

Mineralization the process whereby an organic substance becomes impregnated with, or is converted into, an inorganic substance

Modality any of the various types of sensation, e.g. vision or hearing

Modular made up from independent or standardized units

Modulus of elasticity a measure of the tendency to elastic deformation of a material, also known as YOUNG'S MODULUS

Morphogenesis the biological process that causes an organism to take on its particular shape

Muscular hydrostat a structure found in many animals that is made up almost entirely of muscle fibers and is used to manipulate objects or move the animal around. Examples are the tentacles of octopuses and the trunks of elephants, as well as the tongues of animals

Nanoparticle a structure of diameter between 1 and 100 nanometers (nm)

Network complexity the number of paths and nodes in a network

Neuromast sensory organ in fish consisting of a cluster of receptor cells that runs in a line laterally along the fish's body and is connected with nerve fibers

Node a connection point in a network that is capable of receiving and transmitting information across the network

Osmosis the movement of water molecules across a semipermeable membrane from a solution of lower concentration to one of higher concentration. Osmosis is essential to the passage of water into and out of plant cells

Pathogen a disease-causing biological agent

Phase transition the process whereby a medium changes from one state to another, e.g. solid to liquid

Pheromone a chemical released by an animal to convey a message to another animal, e.g. to mark territory or a pathway, or to indicate readiness for mating

Photonic relating to the generation, transmission, or reception of light

Plasticize to make something pliant or moldable

Polymer a chemical structure made up of a chain of repeated units of smaller molecules. Polymers occur naturally and have been produced synthetically in many forms, and they have a wide range of properties

Polysaccharide a polymer made up of a chain of sugar molecules, associated with energy transfer and storage, and structural functions. Cellulose and chitin are polysaccharides

Potential energy energy that is stored in a system as a result of its physical configuration or some previously applied force, called "potential" because it can be, but has not yet been, converted into some other form of energy and in the process do work

Propagation the spreading, extension, or dissemination of a process or event

Protein a compound made up of amino acids in a linear chain folded into a globular shape. Proteins are essential to organisms, performing mechanical and structural functions and acting in catalytic and metabolic processes

Quorum sensing a decision-making system used by social organisms that relies on the recognition that the number of individuals has reached the threshold that will allow a particular collective function or process to take place

Refractive index a measure of how much the speed of light is reduced in passing through a particular medium. Materials with a high refractive index will slow down light more than those with a lower refractive index

Regolith a layer of loose material that lies over a solid rock layer, e.g. soil, sand, alluvial deposits, gravel, etc.

Scalability the aptitude of a system to be enlarged without loss of efficiency or productivity, of importance when considering whether organic microprocesses can be made to work at a human scale

Sensor a type of transducer that converts a physical property into some kind of energy output for measuring purposes

Setae tiny bristles on various organisms that serve to anchor the organism or to collect food microorganisms from a surrounding fluid

Shear strain a deformation that acts parallel to the surface of a material, where a normal strain acts perpendicular to the surface

Shear stress a stress that is applied parallel to the surface of a material, where a normal stress is applied perpendicular to the surface

Side-chain a branch from the parent molecular chain, e.g. of a polymer

Silica silicon dioxide (SiO_2), the most abundant mineral in the Earth's crust, most commonly found as quartz or sand. Silica is also found in the cell walls of diatoms, a common type of phytoplankton

Size-ordered recruitment a way of using the elements of a network or system in the most efficient manner to perform a particular function, starting with the smallest or weakest units and working up through the levels of strength or size, depending on how much force is required

Steady flow system "steady flow" is a term used in fluid mechanics to describe the movement of a fluid where the velocity or direction of the flow does not vary over time

Stiffness resistance to deformation, e.g. stretching, bending, or compression

Strength resistance to fracture

Stress-strain curve the graph of the relationship between the force (stress) applied to a material and the resultant deformation (strain)

Superorganism an organism that is made up of many organisms, usually applied to social insects such as bees and ants that display a large degree of cooperation and interdependence

Tensile strength resistance to fracture when under tension

Thermosiphoning method of circulatory heat exchange in fluids whereby warmer fluid rises and draws cooler fluid in to replace it, before itself cooling and falling to the bottom of the system once more

Timbre the quality of a musical note or a sound that distinguishes the means of its production, e.g. the characteristic that allows a hearer to distinguish between a trumpet and a trombone when both are playing a note of the same pitch and loudness

Transducer a device that converts one type of energy or physical attribute to another. Sensors and actuators are types of transducer

Turgor the rigidity of plant or animal cells resulting from the presence of a fluid. When plants lose water, they droop as a result of loss of turgor

Ultrasonic relating to sounds of higher frequency than is detectable by human hearing, upward of about 20 kHz

Unsteady flow a term used in fluid mechanics to describe the movement of a fluid where the velocity or direction of the flow varies over time

Valve a device for controlling the passage of fluid through a mechanism, especially one that allows flow in only one direction

Viscoelasticity a property of materials that show both elastic and viscous properties under stress

Viscosity the measure of a fluid's resistance to deformation under stress

Vortex a flow of gas or liquid in a spiral pattern. Vortices can be caused by the passage of a solid body through a fluid, contain a lot of energy, and are slow to dissipate

Young's modulus *see* MODULUS OF ELASTICITY

Contributors

JEANNETTE YEN
Jeannette Yen's Ph.D. is in biological oceanography. She is the Director of Georgia Tech's Center for Biologically Inspired Design and a Professor in the School of Biology. The Center facilitates research into innovative products and techniques based on biologically inspired design solutions that have been concept-tested over millions of years of evolution. Experiencing the benefits of nature as a source of innovative and inspiring principles encourages us to preserve and protect the natural world rather than simply to harvest its products.

YOSEPH BAR-COHEN
Yoseph Bar-Cohen is a Senior Research Scientist at Jet Propulsion Laboratory in Pasadena, California. He received his Ph.D. in Physics (1979) from the Hebrew University, Jerusalem, Israel. He has made two notable discoveries of ultrasonic wave phenomena in composite materials and has coauthored over 320 publications. He challenged scientists worldwide to develop a robotic arm driven by EAP to wrestle with a human and win. In 2003 *Business Week* named him as one of five technology gurus who are "Pushing Tech's Boundaries."

TOMONARI AKAMATSU
Tomonari Akamatsu is an underwater bioacoustician in Japan. He leads research into monitoring endangered marine mammals using passive acoustic techniques. He has a masters degree in physics and a Ph.D. in agriculture, and is a primary investigator of the development of the dolphin mimetic sonar system. He divides some of his time between the editorial board of *Bioinspiration & Biomimetics,* the *Journal of Ethology,* and reviewing the *Journal of the Acoustical Society of America, Marine Ecology Progress Series.*

ROBERT ALLEN

Consultant editor Robert Allen holds a Personal Chair in Biodynamics and Control at the Institute of Sound and Vibration Research (ISVR), University of Southampton, UK. He was awarded a Ph.D. at the University of Leeds in the UK for research on modeling the dynamic characteristics of neural receptors. His research interests focus on the development and application of signal processing techniques for biomedical systems analysis and on the bioinspired control of unmanned underwater vehicles.

STEVEN VOGEL

Steven Vogel is James B. Duke Professor, Emeritus, in the Department of Biology at Duke University, Durham, North Carolina. He received his doctorate at Harvard. While a biologist by training and inclination, he looks at mechanical factors behind the designs of organisms, in particular at their fluid dynamic devices. He has written several books, and articles for a variety of popular magazines. His most recent projects include an undergraduate textbook and a collection of essays on comparative biomechanics.

JULIAN VINCENT

Julian Vincent is Professor of Biomimetics in the Department of Mechanical Engineering at the University of Bath, UK. He has published over 300 papers, articles, and books. His interests cover TRIZ (the Russian system for creative solution of problems), mechanical design of plants and animals, complex fracture mechanics, texture of food, design of composite materials, natural material use in technology, advanced textiles, smart systems, and structures. In 1990 he won the Prince of Wales Environmental Innovation Award.

Bibliography

CHAPTER 1

MARINE BIOLOGY

Aizenberg, Joanna, Alexei Tkachenko, Steve Weiner, Lia Addadi & Gordon Hendler. "Calcitic microlenses as part of the photoreceptor system in brittlestars." *Nature* Vol 412 (August 23, 2001): www.nature.com 819–822.

Aizenberg, Joanna, Vikram C. Sundar, Andrew D. Yablon, James C. Weaver, and Gang Chen. "Biological glass fibers: Correlation between optical and structural properties." (2004): 3358–3363 PNAS March 9, Vol. 101 no. 10.

Aizenberg, Joanna, James C. Weaver, Monica S. Thanawala, Vikram C. Sundar, Daniel E. Morse, Peter Fratzl. "Skeleton of Euplectella sp.: Structural Hierarchy from the Nanoscale to the Macroscale." *Science* Vol 309 (July 8, 2005): 275–278.

Allen, J. J., and A. J. Smits. "Energy Harvesting Eel." *Journal of Fluids and Structures* 15, 1–12.

Arzt, Eduard, Stanislav Gorb, and Ralph Spolenak. *"From micro to nano contacts in biological attachment devices."* PNAS (September 16, 2003): ol. 100 no. 19 10603–10606

Ayers, Joseph, and Jan Witting. "Biomimetic approaches to the control of underwater walking machines." Phil. Trans. R. Soc. A 365 (2007): 273-295

Bartol, I. K. , M. S. Gordon, P. Webb, D. Weihs, and M. Gharib. "Evidence of self-correcting spiral flows in swimming boxfishes." Bioinsp. Biomim. (2008): 3 014001 (7pp) Communication.

Beal, D. N. , F. S. Hover, M. S. Triantafyllou, J. C. Liao and G. V. Lauder. "Passive propulsion in vortex wakes." *Journal of Fluid Mechanics* (2006): 549 : 385-402

Beal, D.N. http://www.bioone.org.www.library.gatech.edu:2048/doi/full/10.1093/ict/42.5.1026 - affl#affl.

Bevan, D. J. M., Elisabeth Rossmanith, Darren K. Mylrea, Sharon E. Ness, Max R. Taylor, and Chris Cuff. "On the structure of aragonite—Lawrence Bragg revisited." *Acta Cryst.* B58 (2002): 448–456.

Capadona, Jeffrey R., Kadhiravan Shanmuganathan, Dustin J. Tyler, Stuart J. Rowan, and Christoph Weder. "Stimuli-Responsive Polymer Nanocomposites Inspired by the Sea Cucumber." *Dermis. Science,* 319 (2008): 1370-1374.

Chan, Brian, N. J. Balmforth, A. E. Hosoi. "Building a better snail: Lubrication and adhesive locomotion." *Physics Of Fluids* 17 (2005): 113101-1, 113101-10.

Chen, P.-Y.; A. Y.-M. Lin; A. G. Stokes; Y. Seki; S. G. Bodde; J. McKittrick, and M. A. Meyers. "Structural Biological Materials: Overview of Current Research." JOM; Jun; 60, 6; ABI/INFORM Trade & Industry (2008): 23-32.

Cohen, Anne L. and Daniel C. McCorkle, Samantha de Putron, Glenn A. Gaetani and Kathryn A. Rose. "Morphological and compositional changes in the skeletons of new coral recruits reared in acidified seawater: Insights into the biomineralization response to ocean acidification." *Geochemistry Geophysics Geosystems* Vol 10, Number 7 (2009): 1–12.

Dabiri, J .O., Colin, S. P., and Costello, J. H. *"Morphological diversity of medusan lineages constrained by animal-fluid interactions." J. Exp. Biol.* 210 (2007): 1868–1873.

Dorgan, Kelly M., Peter A. Jumars, Bruce D. Johnson, and Bernard P. Boudreau. "Macrofaunal Burrowing: The Medium Is The Message." *Oceanography and Marine Biology: An Annual Review,* 44 (2006): 85–121 © R. N. Gibson, R. J. A. Atkinson, and J. D. M. Gordon, Editors. Taylor & Francis.

Dorgan, Kelly M., Peter A. Jumars, Bruce Johnson, B. P. Boudreau, and Eric Landis. "Burrow extension by crack propagation." *Nature* Vol 433 (2005): 475.

Grasso, Frank W., and Pradeep Setlur. "Inspiration, simulation and design for smart robot manipulators from the sucker actuation mechanism of cephalopods." Bioinsp. Biomim. 2 (2007): S170–S181 doi:10.1088/1748-3182/2/4/S06.

Ernst, E. M., B. C. Church, C. S. Gaddis, R. L. Snyder, and K. H. Sandhage. "Enhanced Hydrothermal Conversion of Surfactant-modified Diatom Microshells into Barium Titanate Replicas." J. Mater. Res., 22 [5] (2007): 1121–1127

Farrell, Jay A. , Shuo Pang, and Wei Li. "Chemical Plume Tracing via an Autonomous Underwater Vehicle." Sciences *IEEE Journal Of Oceanic Engineering*, Vol. 30, No. 2 (April, 2007): 428–442.

Fudge, Douglas S., Kenn H. Gardner, V. Trevor Forsyth, Christian Riekel, and John M. Gosline. "The Mechanical Properties of Hydrated Intermediate Filaments: Insights from Hagfish Slime *Threads."* Biophysical Journal Vol. 85 (September, 2003): 02015–2027 2015.

Fudge, Douglas S., T. Winegard, R. H. Ewoldt, D. Beriault, L. Szewciw, and G. H. McKinley. "From ultra-soft slime to hard a-keratins: The many lives of intermediate filaments." *Integrative and Comparative Biology*, Vol. 49, number 1 (2009): 32–39

Honeybee Robotics Spacecraft Mechanisms Corporation. Rock Abrasion Tool. www.honeybeerobotics.com

Hover, F.S., and D. K. P. Yue. "Vorticity Control in Fish-like Propulsion and Maneuvering." *Integrative and Comparative Biology* 42(5):1026–1031. 2002 http://www.bioone.org.www.library.gatech.edu:2048/doi/full/10.1093/ict/42.5.1026 - affl#affl.

Hu, David L., Brian Chan, & John W. M. Bush. "The hydrodynamics of water strider Locomotion." *Nature,* Vol. 424 (August 7, 2003): 663-666.

Hudec, R., L. Veda, L. Pina, A. Inneman, and V. Imon. "Lobster Eye Telescopes as X-ray All–Sky Monitors." *Chin. J. Astron. Astrophys*, Vol. 8 Supplement, (2008): 381–385 (http://www.chjaa.org)

Hultmark, Marcus, Megan Leftwich, and Alexander J. Smits. "Flowfield measurements in the wake of a robotic lamprey." *Exp Fluids* (2007): 683–690.

Kang, Youngjong, Joseph J. Walish, Taras Gorishnyy, and Edwin L. Thomas. "Broad-wavelength-range chemically tunable block-copolymer photonic gels." *Nature Materials* 6 (2007): 957–960.

Kazerounian, Kazem, and Stephany Foley. "Barriers to Creativity in Engineering Education: A Study of Instructors' and Students' Perceptions." *Journal of Mechanical Design.* Vol. 129 (July 2007): 761

Kröger, N., and N. Poulsen. "Diatoms – from cell wall biogenesis to nanotechnology." *Annu. Rev. Genet.* 42 (2008): 83–107.

Lauder G.V., and E.G. Drucker. "Forces, fishes, and fluids: hydrodynamic mechanisms of aquatic locomotion." *News in Physiological Sciences* 17 (2002): 235–240.

Lee, Haeshin, Bruce P. Lee, and Phillip B. Messersmith. "A reversible wet/dry adhesive inspired by mussels and geckos." *Nature* Vol. 448 (July 19, 2007) doi:10.1038/nature05968

CHAPTER 1

MARINE BIOLOGY CONTINUED

Mäthger, Lydia M., and Roger T. Hanlon. *"Malleable skin coloration in cephalopods: selective reflectance, transmission and absorbance of light by chromatophores and iridophores."* *Cell Tissue Res* (2007): 329:179–186.

Manefield, Michael, Thomas Bovbjerg Rasmussen, Morten Henzter, Jens Bo Andersen, Peter Steinberg, Staran Kjelleberg, and Michael Givskov. "Halogenated furanones inhibit quorum sensing through accelerated LuxR turnover." *Microbiology* 148 (Pt 4) (April, 2002): 1119–27 11932456.

McHenry, Matthew. J. "Comparative Biomechanics: The Jellyfish Paradox Resolved." *Current Biology* Vol. 17 No. 16 (2007): R632-R633.

Pelamis Wave Power. http://www.pelamiswave.com/index.php

Peleshanko, Sergiy, Michael D. Julian, Maryna Ornatska, Michael E. McConney, Melbourne C. LeMieux, Nannan Chen, Craig Tucker, Yingchen Yang, Chang Liu, Joseph A. C. Humphrey, and Vladimir V. Tsukruk. "Hydrogel-Encapsulated Microfabricated Haircells Mimicking Fish Cupula Neuromast." *Adv. Mater.* 19 (2007): 2903–2909

Pfeifer, Rolf, Max Lungarella, Fumiya Iida. "Self-Organization, Embodiment, and Biologically Inspired Robotics." *Science* 318 (2007): 1088.

Ralston, Emily, and Geoffrey Swain. "Bioinspiration—the solution for biofouling control?" *Bioinsp. Biomim.* 4 (2009): 015007 (9pp) .

Scardino, Andrew, Rocky De Nys, Odette Ison, Wayne O'Connor, and Peter Steinberg. "Microtopography and antifouling properties of the shell surface of the bivalve molluscs Mytilus Galloprovincialis and Pinctada imbricata." *Biofouling,* 19 Suppl (April, 2003): 221-30 14618724.

Silver J., C. Gilbert, P. Sporer, and A. Foster. "Low vision in east African blind school students: need for optical low vision services." *Br J Ophthalmol* (1995): 79:814–820. http://www.ted.com/talks/josh_silver_demos_adjustable_liquid_filled_eyeglasses.html

Techet. http://www.bioone.org.www.library.gatech.edu:2048/doi/full/10.1093/ict/42.5.1026 - affl#affl.

Triantafyllou. http://www.bioone.org.www.library.gatech.edu:2048/doi/full/10.1093/icb/42.5.1026 - aff1#aff1, M. S., A. H.

Weissburg, M. J., D. B. Dusenbery, H. Ishida, J. Janata, T. Keller, P. J. W. Roberts, D.R. Webster. "A multidisciplinary study of spatial and temporal scales containing information in turbulent chemical plume tracking." *J. Environmental Fluid Mechanics (2002):* 2:65-94.

Winter, Amos G., A. E. Hosoi, Alexander H. Slocum, and Robin L. H. Deits. "The Design And Testing Of Roboclam: A Machine Used To Investigate And Optimize Razor Clam-Inspired Burrowing Mechanisms For Engineering Applications." *Proceedings of the ASME 2009 International Design Engineering Technical Conferences & Computers and Information in Engineering Conference IDETC/CIE 2009.* (August 30 – September 2, 2009): San Diego, California, USA DETC2009-86808.

Yen, J., P. H. Lenz, D. V. Gassie, and D. K. Hartline. "Mechanoreception in marine copepods: Electrophysiological studies on the first antennae." *Journal of Plankton Research* 14 (4) (19920: 495–512.

Yen, J. ,and J. R. Strickler. *"Advertisement and concealment in the plankton: What makes a copepod hydrodynamically conspicuous?" Invert. Biol.*115 (1996): 191–205.

Yeom, Sung-Weon, and Il-Kwon Oh. *"A biomimetic jellyfish robot based on ionic polymer metal composite actuators." Smart Mater. Struct.* 18 (2009): 085002 (10pp) doi:10.1088/0964-1726/18/8/085002.

Zhu, Q. http://www.bioone.org.www.library.gatech.edu:2048/doi/full/10.1093/ict/42.5.1026 - affl#affl.

CHAPTER 2
HUMANLIKE ROBOTS

Abdoullaev A. *Artificial Superintelligence,*, F.I.S. Intelligent Systems, 1999.

Arkin R. *Behavior-Based Robotics.* Cambridge, MA: MIT Press, 1989.

Asimov I. *"Runaround"* (originally published in 1942), reprinted in I Robot, (1942) pp. 33–51.

Asimov I. *I Robot* (a collection of short stories originally published between 1940 and 1950), London: Grafton Books, 1968.

Bar-Cohen, Y., (Ed.). "Proceedings of the SPIE's Electroactive Polymer Actuators and Devices Conf., 6th Smart Structures and Materials Symposium." *SPIE Proc.* Vol. 3669, (1999): pp. 1-414.

Bar-Cohen Y., and C. Breazeal (Eds.). *Biologically-Inspired Intelligent Robots,* 2003. Bellingham, Washington, SPIE Press, Vol. PM122, pp. 1–393.

Bar-Cohen Y. (Ed.). *Electroactive Polymer (EAP) Actuators as Artificial Muscles – Reality, Potential and Challenges,* 2nd Edition, Bellingham, Washington, SPIE Press, 2004, Vol. PM136, pp. 1–765

Bar-Cohen Y., (Ed.). *Biomimetics – Biologically Inspired Technologies,* Boca Raton, FL; CRC Press, 2005, pp. 1–527.

Bar-Cohen Y. and D. Hanson. *The Coming Robot Revolution – Expectations and Fears About Emerging Intelligent, Humanlike Machines,* New York; Springer, 2009.

Breazeal C. *Designing Sociable Robots.* Cambridge, MA: MIT Press, 2002.

Čapek K. *Rossum's Universal Robots (R.U.R.),* Nigel Playfair (Author), P. Selver (Translator), Oxford University Press, USA, 1961.

Dautenhahn K., and C. L. Nehaniv (Eds.). *Imitation in Animals and Artifacts,* Cambridge MA: MIT Press, 2002.

Dietz P. *People are the same as machines – Delusion and Reality of artificial intelligence,* Bühler & Heckel, 2003 (in German).

Drezner T. and Z. Drezner. "Genetic Algorithms: Mimicking Evolution and Natural Selection in Optimization Models," Chapter 5 in [Bar-Cohen, 2005], pp. 157–175.

Fornia A., G. Pioggia, S. Casalini, G. Della Mura, M. L. Sica, M. Ferro, A. Ahluwalia, R. Igliozzi, F. Muratori, and D. De Rossi. "Human-Robot Interaction in Autism." *Proceedings of the IEEE-RAS International Conference on Robotics and Automation (ICRA 2007),* Workshop on Roboethics, Rome, Italy (April 10-14, 2007).

Full R. J., and K. Meijir. "Metrics of Natural Muscle Function," Chapter 3 in [Bar-Cohen, 2004], pp. 73–89.

Gallistel C. *The Organization of Action,* Cambridge, MA. *MIT Press,* 1980.

Gallistel C. *The Organization of Learning,* Cambridge, MA, MIT Press, 1990.

Gates B. "A Robot in Every Home." Feature Articles, *Scientific American* (January, 2007).

Gould, J. *Ethology.* New York: Norton, 1982.

Hanson D. "Converging the Capability of EAP Artificial Muscles and the Requirements of Bio-Inspired Robotics." *Proceedings of the SPIE EAP Actuators and Devices (EAPAD) Conf.,* Y. Bar-Cohen (Ed.), Vol. 5385 (SPIE), (2004): pp. 29-40.

Hanson D. *"Humanizing interfaces—an integrative analysis of the aesthetics of humanlike robots,"* PhD Dissertation, The University of Texas at Dallas: May 2006.

Hanson D. "Robotic Biomimesis of Intelligent Mobility, Manipulation and Expression." Chapter 6 in [Bar-Cohen, 2005], pp. 177–200.

Harris G. "To Be Almost Human Or Not To Be, That Is The Question." *Engineering* Feature Article, (Feb 2007): pp. 37–38.

HUMANLIKE ROBOTS CONTINUED

Hecht-Nielsen R. *"Mechanization of Cognition,"* Chapter 3 in [Bar-Cohen, 2005], pp. 57–128.

Hughes H. C., *Sensory Exotica a World Beyond Human Experience.* Cambridge, MA: MIT Press, 1999. pp. 1–359.

Kerman J. B. *Retrofitting Blade Runner: Issues in Ridley Scott's Blade Runner and Philip K. Dick's Do Androids Dream of Electric Sheep?,* Bowling Green, OH: Bowling Green State University Popular Press, 1991.

Kurzweil R. *The Age of Spiritual Machines: When Computers Exceed Human Intelligence.* New York: Penguin Press, 1999.

Lipson H. "Evolutionary Robotics and Open-Ended Design Automation." Chapter 4 in [Bar-Cohen, 2005], pp. 129–155.

McCartney S. *ENIAC: The Triumphs and Tragedies of the World's First Computer.* New York: Walker & Company, 1999.

Menzel P. and F. D'Aluisio. *Robo sapiens: Evolution of a New Species,* Cambridge, MA: MIT Press, 2000, 240 pages.

Mori M., "The Uncanny Valley." *Energy,* 7(4), (1970): pp. 33–35. (Translated from Japanese to English by K. F. MacDorman and T. Minato).

Perkowitz S., *Digital People: From Bionic Humans to Androids.* Washington, D.C., Joseph Henry Press, 2004.

Plantec P. M., and R. Kurzwell (Foreword). *Virtual Humans: A Build-It-Yourself Kit, Complete With Software and Step-By-Step Instructions.* AMACOM/ American Management Association, 2003.

Raibert M. *"Legged Robots that Balance."* Cambridge, MA: MIT Press, 1986.

Rosheim M. *Robot Evolution: The Development of Anthrobotics.* Hoboken, NJ: Wiley, 1994.

Rosheim, M. "Leonardo's Lost Robot." *Journal of Leonardo Studies & Bibliography of Vinciana,* Vol. IX, Accademia Leonardi Vinci (September 1996): pp. 99–110.

Russell S. J. and P. Norvig. *Artificial Intelligence: A Modern Approach* (2nd Edition). Upper Saddle River, NJ: Prentice Hall, 2003.

Shelde P. *Androids, Humanoids, and Other Science Fiction Monsters: Science and Soul in Science Fiction Films.* New York, NY: New York University Press, 1993.

Shelley M. *Frankenstein.* London: Lackington, Hughes, Harding, Mavor & Jones, 1818.

Turing A.M. "Computing machinery and intelligence." *Mind,* 59, (1950), pp. 433–460.

Vincent J. F. V. "Stealing ideas from nature." *Deployable Structures,* S. Pellegrino (Ed). Vienna: Springer, 2005, pp. 51–58.

CHAPTER 3

UNDERWATER BIOACOUSTICS

Akamatsu, T. Wang, D. Wang, K. and Naito, Y. "Biosonar behaviour of free-ranging porpoises." *Proc. R. Soc. Lond.* B 272 (2005): 797–801.

Au, W .W. L. *The sonar of dolphins.* New York: Springer-Verlag, 1993.

Au, W. W. L., M. C. Hastings. *Principles of Marine Bioacoustics.* New York: Springer, 2008.

Au, W. W. L., A. N. Popper, and R .R. Fay. *Hearing by Whales and Dolphins.* Springer Handbook of Auditory Research. New York: Springer, 2000.

Au, W. W. L. and K. J. Benoit-Bird. "Broadband backscatter from individual Hawaiian mesopelagic boundary community animals with implications for spinner dolphin foraging." *J. Acoust. Soc. Am.* 123 (2008): 2884–2894.

Harley, H. E., E. A. Putman, and H. L. Roitblat. "Bottlenose dolphins perceive object features through echolocation." *Nature* 424 (2003): 667–669.

Jones, B. A., T. K. Stanton, A. C. Lavery, M. P. Johnson, P. T. Madsen, and P. L. Tyack. "Classification of broadband echoes from prey of a foraging Blainville's beaked whale." *J. Acoust. Soc. Am.* 123 (2008): 1753–1762.

Matsuo, I., T. Imaizumi, T. Akamatsu, M. Furusawa, and Y. Nishimori. "Analysis of the temporal structure of fish echoes using the dolphin broadband sonar signal." *J. Acoust. Soc. Am.* 126 (2009): 444–450.

Mitson R. B. *Fisheries Sonar.* Farnham, Surrey, UK: Fishing News Books Ltd., 1984.

Reeder, D. B., J. M. Jech,, and T. K. Stanton. "Broadband acoustic backscatter and high-resolution morphology of fish: Measurement and modeling." *J. Acoust. Soc. Am.* 116 (2004): 747–761.

SIMRAD (multibeam echosounder). http://www.simrad.com.

Sound Metrics Corp. (DIDSON). http://www.soundmetrics.com/index.html

Thomas, J., C. Moss, and Vater, M. *Echolocation in bats and dolphins.* Chicago: University of Chicago Press, 2004.

Urick, R. J. *Principles of Underwater Sound,* 3rd Edition. Los Altos Hills, CA: Peninsula Publishing, 1996.

CHAPTER 4

COOPERATIVE BEHAVIOR

Feder, T. "Statistical physics is for the birds." *Physics Today* (American Institute of Physics) (October 2007): 28–30.

Feng, Z., R. Stansbridge, D. White, A. Wood, and R. Allen. "Subzero III – a low-cost underwater flight vehicle." *Proceedings of the 1st IFAC Workshop on Guidance and Control of Underwater Vehicles.* (April 9–11, 2003): 215–19.

Hargreaves, B. "Guided by nature." *Professional Engineering,* (Institution of Mechanical Engineers), 11 (March, 2009): 29–31.

Holldobler, B. and E. O. Wilson. *The Superorganism.* New York/London: W.W. Norton & Co., 2009.

Hou, Y. and R. Allen. "Intelligent behaviour-based team UUVs cooperation and navigation in a water flow environment." *Ocean Engineering,* 35 (2008): 400–16.

Hou, Y. and R. Allen. "Behaviour-based circle formation control simulation for cooperative UUVs." *Proceedings of the IFAC Workshop NGCUV 2008 'Navigation, Guidance and Control of Underwater Vehicles.* (April 6–10, 2008): Paper No 32, 6pgs.

Nakrani, S. and C. Tovey. "From honeybees to internet servers: biomimicry for distributed management of internet hosting centres." *Bioinspiration & Biomimetics,* 2 (2007): S182–S197.

Olariu, S. and A. Y. Zomaya (Editors). *Handbook of Bioinspired algorithms and applications.* London/New York: Chapman & Hall/CRC, 2006.

Pham, D. T., A. Ghanbarzadeh, E. Koc, S. Otri, S. Rahim, and M. Zaidi. "The bees algorithm – a novel tool for complex optimisation problems." *Proceedings of the Innovative Production Machines & Systems* (July 3–14, 2006): 6pp.

Reynolds, C. http:www.red3d.com/cwr/boids/

Reynolds, C. W. "Flocks, herds, and schools: A distributed behavioral model." *Computer Graphics,* 21(4) (SIGGRAPH '87 Conference Proceedings) (1987): 25–34.

Shao, C. and D. Hristu-Varsakelis, D. "Cooperative optimal control: broadening the reach of bio-inspiration." *Bioinspiration & Biomimetics,* 1 (2006): 1–11.

Tautz, J. *The Buzz about Bees.* Berlin/Heidelberg: Springer-Verlag, 2008.

Vogel, S. Chapter 5 in this book.

CHAPTER 5

MOVING HEAT AND FLUIDS

Schmidt-Nielsen, K. *How Animals Work.* Cambridge, UK: Cambridge University Press, 1972.

Schmidt-Nielsen, K. *Animal Physiology,* 5th edition. Cambridge, UK: Cambridge University Press, 1997.

Turner, J. S. *The Extended Organism.* Cambridge, MA: Harvard University Press, 2000.

Vogel, S. *Cats' Paws and Catapults.* New York: W. W. Norton, 1998.

Vogel, S. *Glimpses of Creatures in Their Physical Worlds.* Princeton, NJ: Princeton University Press, 2009.

CHAPTER 6

NEW MATERIALS

Altshuller, G. *The Innovation Algorithm, TRIZ, Systematic Innovation and Technical Creativity*. Worcester, Mass.: Technical Innovation Center Inc., 1999.

Ashby, M. F. *Materials Selection in Mechanical Design* (3rd edn). Oxford: Elsevier, 1992.

Ashby, M. F., and Y. J. M. Brechet. "Designing hybrid materials." *Acta Materialia* **51** (2003): 5801–21.

Barthlott, W., and C. Neinhuis "Purity of the sacred lotus, or escape from contamination in biological surfaces." *Planta* **202** (1997): 1-8.

Brett, C. T., and K. W. Waldron. *Physiology and Biochemistry of Plant Cell Walls.* London: Chapman & Hall, 1966.

Gordon, J. E. *The New Science of Strong Materials, or Why You Don't Fall Through The Floor.* Harmondsworth: Penguin, 1976.

Lakes, R. S. "Materials with structural hierarchy." *Nature* 361 (1993): 511–515.

Mann, S. *Biomimetic Materials Chemistry*: Hoboken, NJ: Wiley-VCH, 1996.

Mattheck, C. *Design in Nature: Learning from Trees.* Heidelberg: Springer, 1998.

McMahon, T. A. and J. T. Bonner. *On Size and Life*. NY: Freeman, 1983.

Neville, A. C. *Biology of Fibrous Composites; Development Beyond the Cell Membrane.* Cambridge, UK: Cambridge University Press. 1993.

Pollack, G. H. *Cells, Gels and the Engines of Life*. Seattle, WA: Ebner & Sons, 2000.

Shirtcliffe, N. J., G. McHale, M. I. Newton, C. C. Perry, and F. B. Pyatt. "Plastron properties of a superhydrophobic surface." *Applied Physics Letters* **89**, (2006): 104106–2.

Thompson, D. W. *On Growth and Form*. Cambridge, UK: Cambridge University Press, 1959.

Vincent, J. F. V. *Structural Biomaterials*. Princeton: Princeton University Press, 1990.

Vincent, J. F. V. "Survival of the Cheapest." *Materials Today*, (2002): 28–41.

Vincent, J. F. V., O. A. Bogatyreva, N.R. Bogatyrev, A. Bowyer, and A.K. Pahl. "Biomimetics—Its Practice and Theory." *Journal of the Royal Society Interface* **3**, (2006): 471–482.

Vincent, J. F. V. and U. G. K. Wegst. "Design and Mechanical Properties of Insect Cuticle." *Arthropod Structure and Development* **33** (2004): 187–199.

Wainwright, S. A., W. D. Biggs, J. D. Currey, and J. M. Gosline *The Mechanical Design of Organisms*. London: Arnold, 1976.

Index

Acknowledgments

AUTHOR ACKNOWLEDGMENTS

CHAPTER 1, **Marine Biology**, *Jeannette Yen:*
Thanks to Frank Fish for giving me the opportunity to contribute to this book. Marc Weissburg and Michael Helms provided constructive criticism and constant encouragement throughout the writing. I also would like to thank Lorraine Turner for her unending patience and delightful support. This material is based upon work partially supported by the National Science Foundation under Grant No. 0737041, entitled: Biologically Inspired Design: A novel interdisciplinary biology-engineering curriculum,which let me continue to teach biologically inspired design to interdisciplinary classes of creative and innovative undergraduate students at Georgia Tech. The wonder and curiosity expressed by the students helped me to select the best examples for this chapter.

CHAPTER 2, **Humanlike Robots,** *Yoseph Bar-Cohen:*
Yoseph Bar-Cohen would like to acknowledge that some of the research reported in his chapter was conducted at the Jet Propulsion Laboratory (JPL), California Institute of Technology, under a contract with National Aeronautics and Space Administration (NASA).

CHAPTER 3, **Underwater Bioacoustics,**
Tomonari Akamatsu:
Tomonari Akamatsu would like to acknowledge the assistance of the Research and Development Program for New Bioindustry Initiatives of Japan.

CHAPTER 4, **Cooperative Behavior,** *Robert Allen:*
Robert Allen is grateful to various funding bodies, and in particular the EPSRC, for their support of this work and to the many Ph.D. students, postdoctoral research fellows, and academic colleagues in the ISVR, University of Southampton, U.K.

CHAPTER 6, **New Materials and Natural Design,**
Julian Vincent:
Apart from the usual thanks to wife and colleagues, Jim Gordon deserves an especial mention as an uninhibited and wide-ranging thinker who showed me that you don't need maths to be an engineer or a designer. To myself, a biologist, this was a moment of release.

PICTURE CREDITS

The publisher would like to thank the following for their permission to reproduce the images in this book. Every effort has been made to acknowledge the images and their owners; however, we apologize if there are any unintentional errors or omissions.

Tomonari Akamatsu: 86, 184.

Alamy/The Print Collector: 61t; The Art Archive: 167r.

Robert Allen: 185.

The Altshuller Institute for TRIZ Studies: 166.

Martin Ansell: 156t.

Yoseph Bar-Cohen: 46, 51r, 53b, 53t, 56, 59t, 184.

D. J. M. Bevan: 40c.

Bridgeman Art Library/Pinacoteca Capitolina: 48.

Corbis/Manfred Danegger: 7; Lawson Wood: 29b; Steven Kazlowski/Science Faction: 34t; Gilles Podevins: 34b; Mosab Omar: 47b; Bettmann: 49l, 162c; BBC: 49r; Issei Kato: 55l; Car Culture: 62l; Jochen Leubke: 62r; Michael Caronna: 63; Lester Lefkowitz: 83b; Andrew Parkinson: 92t; Christian Hager: 101; Keren Su: 102b; Don Mason: 130t, 130b, 131tc; Jeffery L. Rotman: 149t; Jason Stang: 151; Ralph White: 153t.

Daimler AG: 162bl, 162bc, 162br, 163bl, 163br.

Fotolia: 95r, 99l, 170; Kristian Sekulic: 21; TK Video: 28b; Ian Holland: 31b; Petrafler: 61b; Andrea Zabiello: 90t; Jake Borowski: 90b; Marko Becker: 165tc.

Furuno Electric Co. Ltd.: 71b.

Getty Images/Georgina Douwma: 33; Yoshikazu Tsuno: 60b; Alexander Safonov: 79.

David Hanson: 52

John Huisman/Murdoch University: 39tl.

Iberdrola Renovables: 27.

iStockphoto: 84, 99r, 131tr, 139, 148r; Alexander Potapov: 6; Jamie Carroll: 14l; Jonas Kunzendorf: 20; John M. Chase: 35l; Oliver Anlauf: 40r; Martin Hendriks: 41b; Ivonne Strobel: 42l; Rudi Tapper: 42r; Dejan Sarman: 70; Ju Lee: 78r; Gerald Ulder: 87; James Figlar: 100t; Matthew Scherf: 109; Hung Meng Tan; Chanyut Sribua-rawd: 112; Mark Lundborg: 113b; Frank van de Berg: 116; Sean Randall: 119tr; Jane Norton: 124l; Serghei Velusceac: 128; Valeriy Krisanov: 136b; Juan Moyano: 141b; Arnaud Weisser: 143b; Terrain Scan: 144b; Bettina Ritter: 157t; Lobke Peers: 159cr; Alexei Zaycev: 160t.

J. Bionic Engineering: 171t.

A. P. Jackson/J. F. V. Vincent/R. M. Turner: 148l.

Nick Jewell: 32l.

MIT/Donna Coveney: 31t.

Naoto Honda/Fishing Gear and Method Laboratory/National Research Institute of Fisheries Engineering/Fisheries Research Agency: 80l, 80r.

NASA: 91.

Nature Picture Library/Kim Taylor: 8, 9, 10t, 11, 41t; Doug Allan: 13; Ingo Arndt: 18.

Christopher Neinhuis: 160br.

Richard Palmer/George Lauder: 25; Mashahiro Mori: 47t; Tomonari Akamatsu: 83t, 85tr, 85cr, 85br, 85l; Steven Vogel: 113t, 117t, 118, 123, 125l, 127; Julian Vincent: 168.

Photolibrary/Nick Gordon: 71t; Reinhard Dirscherl: 72.

Photos.com: 10b, 17, 24, 26, 32r, 38, 39br, 54, 60t, 65t, 65b, 68, 69t, 69c, 69b, 75, 77, 78l, 81l, 81r, 82, 92b, 104, 106, 107, 114, 115l, 115r, 117b, 119cl, 119tl, 119cr, 120, 121, 122, 125r, 126, 129, 130b, 131tl, 134t, 137l, 137r, 138t, 140, 142b, 144t, 146, 150b, 154t, 157b, 159bl, 165tl.

Scala Archives/Johann Jakob Schlesinger: 167l.

Science Photo Library: 39cl, 50; Dr. Jeremy Burgess: front cover tl, 37r; Davis Scharf: 30; Andrew Syred: 36r, 37l; Tom Mchugh: 43; Dan Sams: 36l; Victor Habbick Visions: 59b; David Vaughan: 93; Pascal Goetgheluck: 108; Lena Untidt/Bonnier Publications: 103; Sinclair Stammers: 100b; Vaughan Fleming: 124r; Eye of Science: front cover bl, br, 14r, 159cl; Dr. Keith Wheeler: 159tr.

StoLotusan®: 160bl, 161b.

Topfoto/Topham Picturepoint: 55r.

University of Leicester: 32r.

Julian Vincent: 142t, 152, 163, 169l, 169r, 171b, 185.

Steven Vogel: 185.

Michael Watkins: 150, 154l, 154r.

Jeannette Yen: 184.